Basic Electrical and Electronic Tests and Measurements

Michael Braccio

RESTON PUBLISHING COMPANY, INC.
A Prentice-Hall Company
Reston, Virginia

Library of Congress Cataloging in Publication Data
Braccio, Michael.
  Basic electrical and electronic tests and measurements.
  Includes index.
  1. Electric measurements. 2. Electronic measurements. I. Title.
TK275.B67   621.37'2   78-17136
ISBN 0-8359-0589-6

© 1978 by Reston Publishing Company, Inc.
A Prentice-Hall Company
Reston, Virginia 22090

All rights reserved. No part of this book may be reproduced in any way, or by any means, without permission in writing from the publisher.

10 9 8 7 6 5 4 3 2 1

Printed in the United States of America

Contents

Preface, vii

Chapter 1 Development of the Electrical Units 1

1-1 General Considerations, **1**
1-2 Absolute Units and Concrete Standards, **3**
1-3 SI Units, **5**
1-4 Check of DC-Voltage Calibration with Weston cell, **5**
1-5 Check of DC-Voltage Calibration with Mercury Batteries, **7**
1-6 Multimeter Calibrator, **8**
1-7 Linear and Logarithmic Scale Units, **9**
1-8 Digital Display, **13**
1-9 Calculator-Aided Computations, **15**
1-10 Basic Multimeter Instrument Circuitry, **17**
1-11 Fundamental Test and Measurement Techniques, **19**
1-12 Instruments Utilized in Various Frequency Ranges, **20**
Review Questions, **23**

iv CONTENTS

Chapter 2   **Statistics and Measurement Errors**   25

       2-1   General Considerations, **25**
       2-2   Precision and Accuracy Considerations, **27**
       2-2   Precision and Accuracy Considerations **27**
       2-3   Voltmeter Sensitivity, **32**
       2-4   Error Computation, **36**
       2-5   Ohmmeter Scale Resolution, **40**
       Review Questions, **42**

Chapter 3   **Resistance Measurements**   43

       3-1   General Considerations, **43**
       3-2   Semiconductor Junction Resistance, **44**
       3-3   Measurement of Nonlinear Resistance Values, **49**
       3-4   Positive Resistance and Negative Resistance, **51**
       3-5   Thermistor Resistance Measurements, **52**
       3-6   Light Dependent Resistor (LDR) Resistance Measurement, **53**
       3-7   Internal, Input, and Output Resistance Measurements, **54**
       3-8   Resistance Measurement with Megohmmeter, **56**
       3-9   Resistance Strain Gage, **57**
       3-10   Wheatstone Bridge Resistance Measurements, **58**
       3-11   Measurement of Battery Internal Resistance, **59**
       3-12   Typical Junction Resistance Values, **60**
       3-13   In-Circuit Measurement Expedient, **60**
       Review Questions, **63**

Chapter 4   **DC Voltage Measurements**   65

       4-1   General Considerations, **65**
       4-2   DC Voltage Distribution in Transistor Circuitry, **68**
       4-3   DC Voltage Changes From Transistor Defects, **77**

CONTENTS   v

          4-4   Tolerances on DC Voltage Values, **82**
          4-5   Measurement of High DC Voltage, **83**
          Review Questions, **85**

Chapter 5   AC Voltage Measurements   87

          5-1   General Considerations, **87**
          5-2   Measurement of AC Voltage with DC Component Present, **89**
          5-3   AC-to-DC Converters, **90**
          5-4   Turnover Error in AC Voltage Measurement, **91**
          5-5   Measurement of Small AC Voltages, **93**
          5-6   Tuned AC Voltmeters, **94**
          5-7   True rms AC Voltage Measurement, **95**
          5-8   Decibel Measurements with AC Voltmeters, **96**
          5-9   Measurement of Negative Impedance, **104**
          Review Questions, **106**

Chapter 6   Oscilloscope Tests and Measurements   107

          6-1   General Considerations, **107**
          6-2   Rise-time Measurement and Transient Response, **110**
          6-3   Circuit Loading by Oscilloscope, **112**
          6-4   Pulse Display on High-Speed Sweep, **113**
          6-5   Measurement of Time Constant, **117**
          6-6   Measurement of Inductor Q Value, **119**
          6-7   Phase Measurement, **121**
          6-8   Linearity Measurement, **124**
          6-9   High Frequency Measurements, **124**
          6-10   Transistor Measurements, **131**
          6-11   Television Network Tests and Measurements, **135**
          6-12   TV Receiver Waveform Characteristics, **147**
          6-13   Ignition Waveform Analysis, **153**
          Review Questions, **163**

Chapter 7 Audio Measurements 165

7-1 General Considerations, **165**
7-2 Measurement of Frequency Response, **166**
7-3 Measurement of Harmonic and Intermodulation Distortion, **172**
7-4 Measurement of Music Power Capability, **177**
7-5 Measurement of Impedance, Inductance, and Capacitance Values, **179**
7-6 Stereo Decoder Separation Measurement, **183**
7-7 Audio Units, **184**
7-8 Tone-Burst Test of Speaker Enclosure, **187**
Review Questions, **190**

Chapter 8 Digital Measurements 191

8-1 General Considerations, **191**
8-2 Digital Signal Characteristics, **192**
8-3 Digital Probe Application Techniques, **196**
8-4 Oscilloscope Waveform Analysis, **200**
8-5 Digital Delay Techniques, **206**
8-6 Microprocessor Tests and Measurements, **208**
Review Questions, **231**

Appendix i Resistor Color Code, 233

Appendix ii Capacitor Color Code, 235

Appendix iii Basic Electrical Circuit Equations, 237

Index, 241

Preface

With the rapid advance of electronics technology, measurements are becoming increasingly important. The advent of digital technology has greatly increased the accuracy that can be realized in routine measurements. It is essential for the student to acquire a basic grasp of the science and art of measurements, starting with the basic processes. He should understand how the standards of measurement are derived and applied to technological procedures, and the accuracy that may be expected from them. A study of measuring processes is directly related to electrical circuit and field theory, and it serves as a review and an integration of many branches of the discipline. Electrical units originate in theoretical definitions and do not become available for use until they have been realized by a measuring process. Students should be thoroughly prepared for this broad field; some understanding should be acquired of basic high-precision methods, but not to the exclusion of less precise but more commonly used methods.

Development of the electrical units is considered in the first chapter, with a preliminary introduction to measuring techniques. The second chapter explains the elements of statistics and the sources of measurement errors. Basic distinction is made between accuracy and precision concepts, with discussion of instrument errors and observational errors. Resistance measurements are detailed in the third chapter, with attention to both linear and nonlinear circuit parameters. Practical approaches to the measurement of internal, output, and input resistance values are explained. In the fourth chapter, DC voltage measurements are covered. Students are sometimes confused by electron versus hole flow concepts, and this distinction is carefully illustrated. Tolerances on DC voltage values are explained, and changes in DC-voltage distribution owing to variations in grounding points

viii PREFACE

are exemplified. DC-voltage changes owing to transistor defects are also noted.

AC voltage measurements are discussed in the fifth chapter. Instrument requirements are explained and sources of error are enumerated. Instruments for the measurement of small AC voltages and tuned AC voltmeters are included in the text. Because technicians and engineers generally measure decibel values with AC voltmeters, correct techniques are detailed for these procedures. Waveform errors are pointed out and the practical management of nonsinusoidal waveforms is introduced. Oscilloscope tests and measurements are covered in the sixth chapter. Basic measurements of rise time, transient response, time constants, Q values, phase, and amplitude linearity are explained and illustrated. Fundamental waveform analysis is included. Essential television network tests and measurements with the vertical-interval test signal are introduced.

Audio measurements are considered in the seventh chapter. Frequency response, harmonic and intermodulation distortion, music-power capability, and impedance, inductance, and capacitance measurements are detailed. Stereo decoder separation measurements are included. A discussion of specialized audio units, such as the phon and the mel, is presented. In the eighth chapter, digital tests and measurements are discussed and illustrated. Digital signal characteristics are explained, with an introduction to the data domain versus the time-frequency domain. Operation of logic state analyzers is discussed as a preliminary procedure to detailed pulse analysis with the oscilloscope. Microprocessor tests and measurements are explained with specific consideration of the widely used 4004 and 8008 types.

The author is indebted to various manufacturers for illustrative and technical material, as noted throughout the text. He is particularly grateful to the Hewlett-Packard Company for their welcome cooperation in supplying state-of-the-art microprocessing test data. In addition, the author wishes to take this opportunity to thank his fellow instructors for their suggestions and constructive criticisms. This text is intended primarily as a teaching tool in our junior colleges, technical institutes, and vocational schools. It is also designed as an appropriate guidebook for the home-study student. In addition, on-the-job technicians and junior engineers will find this work a useful handbook for ready reference.

chapter one

Development of the Electrical Units

1-1 General Considerations

Accurate electrical measurements are as essential to the electrical engineer or technician as an accurate tape line to the surveyor or an accurate time source to the navigator. As in any other technical field, progress depends on units of measurement that are reliable, accurate, and available to workers everywhere. Electrical scientists have gradually developed a system of such units to a refinement far removed from the crude beginnings of early investigators. Measuring instruments have simultaneously undergone extensive development. As an illustration, the digital voltmeter shown in Figure 1-1 provides considerably higher indication accuracy than the prior generation of analog-type voltmeters.

The early stages of electrical science from 550 B.C. to 1700 A.D. were characterized, as would be expected, by qualitative observations of electrical phenomena. In following centuries, the nature of electrical actions became better understood, and quantitative relationships were gradually established. The first electrical experiments were made by rubbing glass rods with silk, and observations were made of attraction and repulsion of unlike and like charges. From these early experiments, the gold-leaf *electrometer* was developed; this was the first voltmeter used by scientists. Although the electrometer was insensitive in comparison with present-day voltage-measuring instru-

2 DEVELOPMENT OF THE ELECTRICAL UNITS

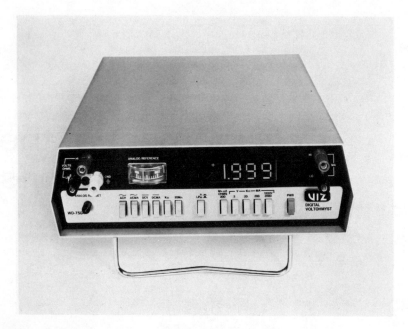

Figure 1-1. A highly accurate digital voltmeter. (*Courtesy,* VIZ)

ments, it opened a new era of quantitative observations of electrical phenomena.

Production of a steady flow of electricity became available with the invention of the voltaic pile by Volta in 1800. In turn, the quantitative investigation of electrical circuits started in 1827, when George Simon Ohm discovered the relationship or "law" that bears his name. Note that volt, ampere, and ohm units were not established until some years later. *Relative* current values were measured with a compass-needle and coil arrangement (tangent galvanometer). Voltage values were stated in terms of the potential supplied by a voltaic pile of specified construction. Resistance values were stated in terms of the resistance of a particular length of iron or copper wire with a diameter specified by the individual experimenter. By way of comparison, scientists, engineers, and technicians today employ high-precision standard resistors for instrument calibration (see Figure 1-2).

It became evident that a universal system of units was needed by all workers in the electrical field, and that these electrical units should be related to the established mechanical units of length, force, mass, and time. In 1832, Karl Friedrich Gauss measured the strength of the earth's magnetic field in terms of length, mass, and time. In 1849, Friedrich Wilhelm Georg Kohlraush measured resistance in terms of these units. Wilhelm Eduard Weber, in 1851, introduced a complete system of electrical units that were

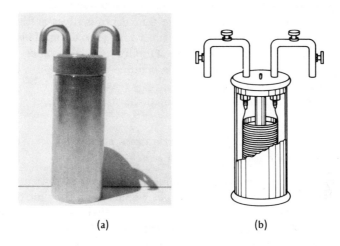

Figure 1-2. High-precision standard resistor: (a) appearance; (b) construction.

based on the mechanical units. Weber's principles form the basis of our present system of electrical measurements. In 1861, the British Association for the Advancement of Science started work on specification of the standard ohm, and announced the British Association unit in 1864. This consisted of a specified coil of platinum-silver alloy wire sealed in a container filled with paraffin. This standard continued in use for twenty years.

1-2 Absolute Units and Concrete Standards

Distinction is made between a theoretically defined electrical unit and a standard that is used in measurement procedures. For example, an ohm is defined in accordance with a logical system of electrical units. On the other hand, a resistance standard is designed for direct and convenient use in laboratory measuring arrangements. Otherwise stated, concrete working standards of resistance are utilized to determine the values of resistors which may be brought to the laboratory for calibration. The duty of maintaining the most precise standards (*primary* standards) is usually assigned to the national laboratories. All *secondary* standards, such as used in industrial laboratories, are calibrated against primary standards. The standard resistor illustrated in Figure 1-2 is an example of a secondary standard.

As noted previously, early determination of the electric current unit was made by means of a tangent galvanometer with respect to the earth's magnetic field. This method was abandoned subsequently owing to variations in the intensity of the earth's magnetic field, and because of disturbances from electric power lines and installations. Therefore, the interna-

4 DEVELOPMENT OF THE ELECTRICAL UNITS

tional ampere was redefined in 1900 by an electrolytic standard by the London Conference as the current value that deposits silver at the rate of 0.00111800 grams/sec from a standard silver-nitrate solution. This electrolytic standard represented a primary standard. On the other hand, a calibrated (analog) current meter such as pictured in Figure 1-3 is termed a secondary standard.

The London Conference then defined the unit of electromotive force (voltage) as follows:

An international volt is the potential difference produced by an international ampere flowing through an international ohm.

However, the concept of a voltaic cell of some sort as a concrete standard of voltage had occurred to various workers from the time of Volta's invention. The Daniell cell (after Daniell) was first used as a concrete standard, but was subsequently abandoned because of its short life. Finally, in 1893, Weston produced a cadmium cell that has made a remarkable record for constancy of electromotive force (EMF) over long periods of time. An International Technical Committee that met in Washington in 1910 established the value of 1.01830 V for the EMF of the standard Weston cell. This cell employs a saturated cadmium-sulphate solution. A cell with an unsaturated cadmium-sulphate solution, however, is often preferred for shop

Figure 1-3. An analog-type current meter. (*Courtesy,* Simpson Electric Co.)

1-4 CHECK OF DC-VOLTAGE CALIBRATION WITH WESTON CELL

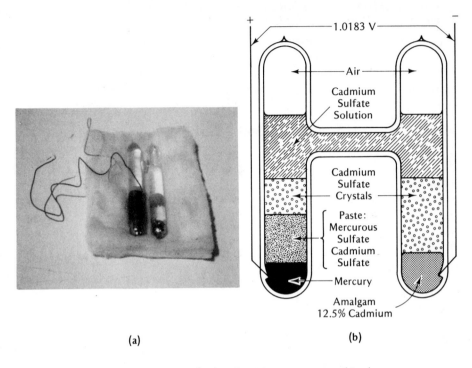

Figure 1-4 Weston-type standard cells: (a) appearance; (b) plan of a saturated cell.

use because of its very small temperature coefficient. An unsaturated cell, often termed a *shop cell* (Figure 1-4) is classed as a secondary standard.

1-3 SI Units

The International System of Units, or SI System, is derived from the metric system of physical units and denotes the French Système Internationale d'Unites. It was adopted, defined, and named in 1960 by the General Conference on Weights and Measures. The IEEE standards designate letter symbols for electrical engineering quantities in the SI System as listed in Table 1-1. Four of the basic units in the SI System are length, mass, time, and electric current.

1-4 Check of DC-voltage Calibration with Weston cell

High-accuracy dc voltmeters such as illustrated in Figure 1-1 are often checked with a Weston shop cell. A shop cell provides a terminal voltage from 1.0185 to 1.0190 V at 20°C. Although its accuracy is not as great as that of a saturated (normal) cell, a shop cell is adequate for checking the

Table 1-1
Letter Symbols for Quantities and Units in Electrical Engineering Designated in the SI System

Quantity	Quantity Symbol	SI Unit	Unit Symbol	Identical Unit
charge	Q	coulomb	C	A · s
current	I	ampere	A	
voltage	$V, E \ldots U$	volt	V	W/A
electromotive force	V	volt	V	
potential difference	V, ϕ	volt	V	
resistance	R	ohm	Ω	V/A
conductance	G	siemen	S	A/V
reactance	X	ohm	Ω	V/A
susceptance	B	siemen	S	A/V
impedance	Z	ohm	Ω	V/A
admittance	Y	siemen	S	A/V
capacitance	farad	farad	F	C/V
inductance	L	henry	H	Wb/A
energy, work	W	joule	J	N · m
power (active)	P	watt	W	J/s
power · apparent	$S \ldots P$	voltampere	VA	
power · reactive	$Q \ldots P^s q$	var	var	
resistivity	ρ	ohm · meter	Ω · m	
conductivity	γ, σ	siemen per meter	S/m	
electric flux	ψ	coulomb	C	
electric flux density, displacement	D	coulomb per square meter	C/m^2	
electric field strength	E	volt per meter	V/m	
permittivity	ε	farad per meter	F/m	
relative permittivity	ε, κ	(numeric)		
magnetic flux	ϕ	weber	Wb	V · s
magnetomotive force	$F \ldots \mathcal{F}$	ampere (amp turn)		
reluctance	$R \ldots \mathcal{R}$	ampere per weber	A/Wb	
		reciprocal henry	H^{-1}	
permeance	$P \ldots \mathcal{P}$	weber per ampere	Wb/A	
		henry	H	
magnetic flux density	B	tesla	T	Wb/m^2
magnetic field strength	H	ampere per meter	A/m	
permeability (absolute)	μ	henry per meter	H/m	
relative permeability	μ_r	(numeric)		

1-5 CHECK OF DC-VOLTAGE CALIBRATION WITH MERCURY BATTERIES

accuracy of any service-type voltmeter. One practical advantage of a shop cell is that it does not have to be used with an elaborate potentiometric calibrating arrangement. In other words, a dc voltmeter can be connected directly to a shop cell, provided only that the current demand is not greater than 10 microamperes (μA). For example, the digital multimeter shown in Figure 1-1 has an input resistance of 10 megohms on its dc-voltage ranges. Accordingly, this meter will draw approximately 0.1 μA from a shop cell. Therefore, it is permissible to connect the test leads of the digital voltmeter directly to the terminals of the shop cell.

1-5 Check of DC-Voltage Calibration with Mercury Batteries

Low-sensitivity voltmeters draw appreciable current. For example, a 1000 ohm-per-volt multimeter operated on its 2.5-V range would draw approximately 0.4 milliampere (mA) from a shop cell. This is approximately four times the maximum tolerable current drain from the cell. Therefore, some other method must be utilized to check the calibration of a low-sensitivity voltmeter. When extremely high accuracy is not required, a mercury battery (Figure 1-5) is a satisfactory source of calibrating voltage. A mercury battery has a reasonably precise terminal voltage until its useful life is ended. This terminal voltage has a higher accuracy than the rated accuracy of typical service-type voltmeters. The mercury battery pictured in Figure 1-5 has

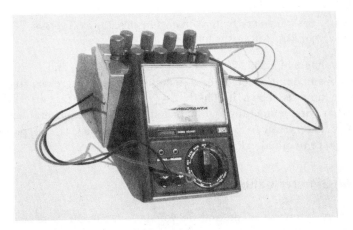

Figure 1-5. A shop-assembled mercury battery checks the indication accuracy of a 1000 ohm-per-volt multimeter on its DC-voltage ranges.

8 DEVELOPMENT OF THE ELECTRICAL UNITS

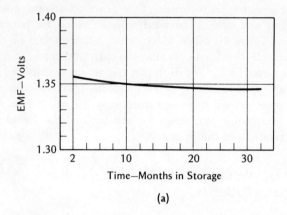

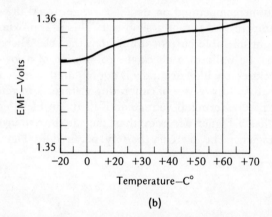

Figure 1-6. Basic mercury battery characteristics: (a) EMF versus age in storage; (b) EMF versus temperature.

output voltages of 1.35, 2.70, 4.05, 5.40, 6.75, 8.10, 9.45, and 10.80 volts. The EMF (open-circuit voltage) of a mercury battery decreases slightly with shelf-life time as shown in Figure 1-6(a). Note also that its EMF varies somewhat with temperature as depicted in (b). As a general summary, a mercury battery can be regarded to have an accuracy of $\pm \frac{1}{2}$ percent with respect to its nominal value.

1-6 Multimeter Calibrator

Secondary standards of dc and ac voltage, dc current, and resistance are utilized to calibrate multimeters. Most multimeters are designed for service application and have moderate accuracy. A typical secondary standard for

1-7 LINEAR AND LOGARITHMIC SCALE UNITS   9

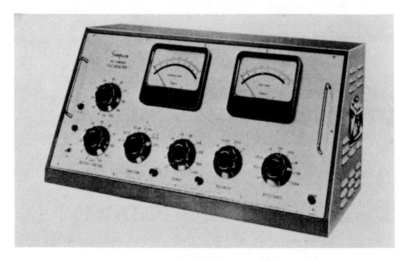

Figure 1-7. A voltage-current-resistance calibrator for checking multimeter accuracy indication. (*Courtesy,* Simpson Electric Co.)

this class of instruments is illustrated in Figure 1-7. Its rated accuracy on dc voltage and current ranges is ±0.5 percent of full scale. Rated accuracy on ac voltage ranges is ±0.75 percent of full scale. Rated accuracy on resistance ranges is ±1 percent. The resistance standards in the calibrator consist of precision wirewound resistors fabricated from special alloys to minimize their temperature coefficient of resistance.

1-7 Linear and Logarithmic Scale Units

Analog meters such as multimeters generally employ both linear and logarithmic scales. As an illustration, the scale plate shown in Figure 1-8 has a linear dc voltage scale and a logarithmic dB scale. Its resistance scale is nonlinear, and its first ac voltage scale is somewhat nonlinear. Before *voltmeter accuracy* factors can be considered, the reader should acquire familiarity with the various meter scales that are available. Decibel units and voltage units are complementary terms. Otherwise stated, if a linear voltage scale is utilized, the decibel scale on the same scale plate will employ logarithmic units. On the other hand, if a linear decibel scale is used, the voltage scale on the same scale plate will be marked in logarithmic units. The Hewlett-Packard Company notes as follows:

> If an instrument has a linear dB scale and a nonlinear voltage scale, it is said to have a "linear-log" scale.

10 DEVELOPMENT OF THE ELECTRICAL UNITS

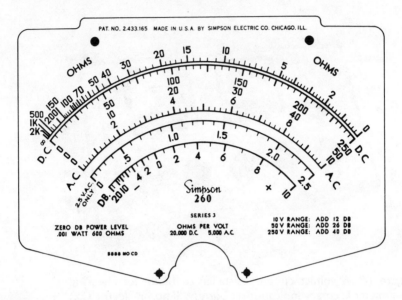

Figure 1-8. A scale plate for a volt-ohm-milliammeter. (*Courtesy, Simpson Electric Co.*)

Various scales provided by HP voltmeters are exemplified in Figure 1-9. Analog meters usually have nonlinearities and/or offsets inherent in attenuators and amplifiers. (refer to Figure 1-10.) Moreover, the meter movement itself can have nonlinearities, despite the fact that the scales may have been individually calibrated. Nonlinearity factors cause *percent-of-reading* errors, and offset factors cause *percent-of-full-scale* errors. Note that percent-of-reading errors are constant, regardless of the pointer position on the scale. On the other hand, percent-of-full-scale error increases in terms of percent-of-reading value as the pointer is positioned farther down the scale.

Insofar as instrument specification sheets are concerned, accuracy specifications are usually expressed in one of three ways: (1) percent of the full-scale value, (2) percent of the reading, and (3) percent of reading plus percent of full-scale value. The first specification is most widely used. The second specification is more commonly applied to meters that have a logarithmic scale. The third specification has been used in more recent practice to provide a tighter accuracy specification on instruments that have linear scales. For a thorough evaluation of instrument accuracy, the following questions should be considered:

1. Does the accuracy rating apply at all input-voltage levels up to the maximum overrange point? (This consideration may be qualified by means of linearity specifications.)

1-7 LINEAR AND LOGARITHMIC SCALE UNITS 11

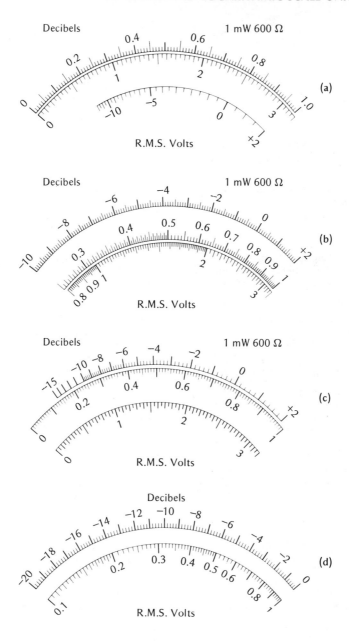

Figure 1-9. Four different types of meter scales provided by HP (Hewlett-Packard) voltmeters: (a) linear 0-3 and 0-10 V scales plus a dB scale: (b) linear dB scale plus nonlinear (logarithmic) voltage scales; (c) dB scale placed on larger arc for greater resolution; (d) linear −20 to 0 dB scale, useful for acoustical and communications applications. (*Courtesy,* Hewlett-Packard)

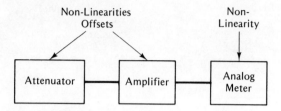

Figure 1-10. Nonlinearities cause percent-of-reading errors; offset causes percent-of-full-scale errors.

Table 1-2
Specifications for a Solid-State Multimeter

DC Volts
10 Ranges: 0–50, 150, 500 mV; 1.5, 5, 15, 50, 150, 500, 1500 V. Accuracy: ±1.5% F.S. Input Resistance: 15 megs, including 100 K in PR-21 probe. AC Rejection: Greater than 46 dB at 60 Hz.

AC Volts
10 RMS Ranges: 0–50, 150, 500 mV; 1.5, 5, 15, 50, 150, 1500 V. 10 Peak-to-Peak Ranges: 0–140, 440 mV; 1.4, 4.4, 14, 140, 440, 1400, 4400 V. Accuracy: ±3% F.S. at 60 Hz; ±5% on 500 V and 1500 V ranges. Input Impedance: 15 megs shunted by 41 pF at input jacks. Frequency Response: 50 mV–150 V ranges, ±0.5 dB, 20 Hz–500 kHz; ±3 dB, 5 Hz–750 kHz.

DC Current
10 Ranges: 0–50, 150, 500 μA; 1.5, 5, 15, 50, 150, 500 mA; 1.5 A. Accuracy: ±3% F.S., except ±4% on 1.5 A range. Internal Voltage Drop: At input terminals, 50 mV to 50 mA range.

AC Current
10 Ranges: 0–50, 150, 500 μA; 1.5, 5, 15, 50, 150, 500 mA; 1.5. Accuracy: ±4% F.S. at 60 Hz; ±5% on 50 μA and 1.5 A ranges. Frequency Response: ±0.5 dB, 20 Hz–5 kHz; ±3 dB, 7 Hz–16 kHz. Internal Voltage Drop: At input terminals, 50 mV to 50 mA range.

Resistance
8 Low-Power Ranges (33 mV source): Rx0.1, Rx1, Rx10, Rx100, Rx1K, Rx10K, Rx100K, Rx1 meg. Center Scale Reading: 1 ohm, Rx0.1; 10 ohms, Rx1 range. Accuracy: ±3° of arc.

Decibels
10 Ranges: −40 to +66 dB. Accuracy: ±3% F.S. (0 dB = 1 mW across 600 ohms).

General
Meter: 7″, with 6.63″ scale arc length; 100 μA movement, ±2%, 100°; mirrored scale. Protection: Meter and FET input protected against overloads. Circuit overload protection by diodes and fuse; spark gaps for high voltage protection. Power Required: 105-124 VAC, 50–60 Hz; also available for 230 VAC, 50/60 Hz. 3.9 W, three-wire grounded cord. Size (HWD): 18 x 20 x 9 cm (7.25 x 8 x 3.63″). Net Weight: 1.96 kg (4.25 lbs.).

Courtesy, B & K Precision, Div. of Dynascan Corp.

2. Does the accuracy rating apply on all ranges of the instrument?
3. Does the accuracy rating apply over a useful temperature range?
4. If temperature variations are not included in the accuracy rating, is the temperature coefficient specified?

General specifications for a typical solid-state multimeter are shown in Table 1-2.

When dc measurements are of primary concern, it is advisable to select an instrument that has the broadest capability in meeting the desired applications. If ac measurements involving sine waves with only moderate amounts of distortion ($< 10\%$) are of primary concern, a voltmeter with average response can perform over a bandwidth up to several megahertz. However, in the case of high-frequency measurements (> 10 MHz), a peak-responding voltmeter with a diode-probe input is the most economical choice. Note that peak-responding instrument circuits are satisfactory if inaccuracies caused by distortion of the input waveform can be tolerated. When measurements are required to determine the effective power of nonsinusoidal waveforms, a true-rms-responding voltmeter is the best choice. As a general rule, true-rms (root-mean-square) meters can indicate the rms values of ac voltages only, inasmuch as they usually employ ac-coupled circuitry. Most true-rms voltmeters have a frequency cutoff point in the vicinity of 20 Hz. This restriction prevents a true-rms voltmeter from indicating the value of any subsonic dc voltage or the value of a dc component in a waveform.

1-8 Digital Display

Basic considerations of digital displays are noted by the Hewlett-Packard Company as follows:

> Digital voltmeters are classified according to the number of full digits displayed. An overrange digit is an extra digit that is added to allow the operator to read beyond full scale. This overrange digit is often termed a "one-half" or a "partial" digit because it cannot display all numbers through 9. Overranging greatly extends a DVM's usefulness by maintaining resolution up to, and beyond, full scale. As an illustration, if a signal changes from 9.999 V to 10.012 V, a 4-digit DVM without overranging could measure the first voltage as "9.999 V," but would require range-switching to make the second measurement, with a resulting change of "10.01 V." This 0.002-V change would not be seen. With overranging provided, the second measurement could be made as "10.012 V " with no loss of resolution.

An example of overranging on a $3\frac{1}{2}$ digit DVM is shown in Figure 1-11 (see also Table 1-3).

14 DEVELOPMENT OF THE ELECTRICAL UNITS

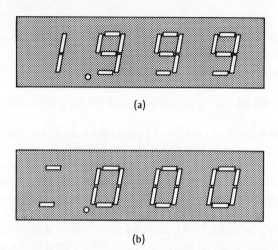

Figure 1-11. Example of $3\frac{1}{2}$ digit displays: (a) if the 1-volt range is selected, the display will indicate up to 1.999 volts; (b) if the input voltage exceeds 1.999 volts, the top and bottom segments of the first digit flash to indicate an out-of-range condition. (*Courtesy*, B & K Precision, Div. of Dynascan Corp.)

Overranging is specified as a percentage. A 4-digit DVM with 100 percent overranging would have a maximum display of "19999." A specification of 20 percent overranging would provide a maximum resolution of "1199." Resolution is the ratio of the maximum number of counts that can be displayed to the least number of counts. Full-scale resolution of a 5-digit DVM is 100,000-to-1, or 0.001 percent. Overranging is generally ignored in resolution. A sensitivity specification denotes the smallest incremental voltage change that the DVM is able to detect. Mathematically, it is the lowest full-scale range multiplied by the resolution of the DVM. Sensitivity of a 5-digit DVM with resolution of 0.001 percent and a 100-mV lowest full-scale range is equal to $0.001\% \times 100$ mV = 1 μV.

An accuracy specification denotes the exactness to which a voltage value can be determined, relative to the legal volt maintained by the U.S. National Bureau of Standards. Accuracy specification is based on errors involved in traceability to N.B.S. as well as errors incurred by the instrument. To be meaningful, accuracy must be stated along with the conditions under which it will hold. These conditions include time, temperature, line variations, and humidity. Conditions specified should be realistic relative to intended application. For example, a DVM specified with a temperature range of $25°C \pm 1°C$ would require a highly controlled environment, whereas $\pm 5°C$ would cover the majority of environments. The period of time over which accuracy holds is particularly important because it indicates

Table 1-3
Measurements and Interpretation of Digital Displays

To Measure		Set Function Switch	Set Range Switch	Connect + Lead	Digital Display	
					Full Scale	100% Overrange
DC Volts	0–1 V	DCV	1 V	(+) Jack	1.000	1.999
	0–10 V		10 V	(+) Jack	10.00	19.99
	0–100 V		100 V	(+) Jack	100.0	199.0
	0–1000 V		1000 V	(+) Jack	1000.	1500.*
AC Volts	0–1 V	ACV	1 V	(+) Jack	1.000	1.999
	0–10 V		10 V	(+) Jack	10.00	19.99
	0–100 V		100 V	(+) Jack	100.0	199.9
	0–1000 V		1000 V	(+) Jack	1000.	1000.**
DC Current	0–1 mA	DCA	1 mA	(+) Jack	1.000	1.999
	0–10 mA		10 mA	(+) Jack	10.00	19.99
	0–100 mA		100 mA	(+) Jack	100.0	199.9
	0–1 Amp		1000 mA	1 A	1000.	1999.
AC Current	0–1 mA	ACA	1 mA	(+) Jack	1.000	1.999
	0–10 mA		10 mA	(+) Jack	10.00	19.99
	0–100 mA		100 mA	(+) Jack	100.0	199.9
	0–1 Amp		1000 mA	1 A	1000.	1999.
Ohms	0–100 Ω	Ohms	100 Ω	(+) Jack	100.0	199.9
	0–1000 Ω		1 KΩ	(+) Jack	1.000	1.999
	0–10 KΩ		10 KΩ	(+) Jack	10.00	19.99
	0–100 KΩ		100 KΩ	(+) Jack	100.0	199.0
	0–1 MΩ		1000 KΩ	(+) Jack	1000.	1999.
	0–10 MΩ		10 MΩ	(+) Jack	10.00	19.99

Courtesy, B & K Precision, Div. of Dynascan Corp.

the DVM's stability and how often it will have to be calibrated. Accuracy is usually expressed as a percentage of the reading plus a percentage of the range (or full scale). Accuracy is always better at or above full-scale indication.

1-9 Calculator-Aided Computations

Various computations must be made in the course of measurement evaluations. Although the associated equations can be solved with pencil and paper, or with a slide rule supplemented by a scratch pad, a pocket digital

16 DEVELOPMENT OF THE ELECTRICAL UNITS

calculator can save appreciable labor. A digital calculator also provides more accurate answers than a slide rule. Engineering-type calculators are comparatively sophisticated, and provide functions that often save many intermediate steps in solution of an equation. Consider the keyboard for the calculator depicted in Figure 1-12. Each trigonometric and logarithmic function, square root, addition, subtraction, multiplication, division, and several other mathematical operations can be performed with a single keystroke.

This type of digital calculator includes an "operational stack" of four registers, plus a data storage register for constants. The stack holds intermediate answers and, at the appropriate time, automatically brings them back for further use. This capability eliminates the need for scratch pads, and the necessity for re-entering intermediate answers when performing chains of calculations such as sums of products or products of sums. In turn, the operator can solve almost any measurement calculation faster, more accurately, and with less labor than with a slide rule or pencil-and-paper.

(a)

Figure 1-12. Digital calculators: (a) appearance of desktop calculator (*Courtesy*, Heath Co.); (b) keyboard of the Hewlett-Packard H-35 pocket calculator. (*Courtesy*, Hewlett-Packard)

1-10 BASIC MULTIMETER INSTRUMENT CIRCUITRY   17

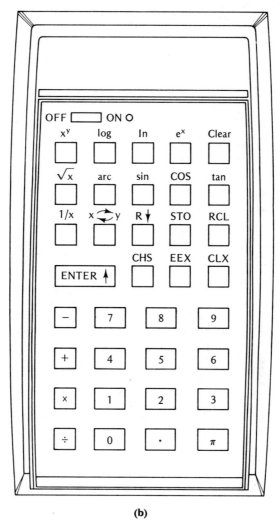

(b)

Figure 1-12. (Cont.)

1-10 Basic Multimeter Instrument Circuitry

A multimeter comprises several instruments, or instrument circuits, within a single case, and is designed to measure at least two electrical quantities. A typical multimeter measures dc voltage, resistance, dc current, and ac voltage. As seen in Figure 1-13, a meter movement can be used either as a

18 DEVELOPMENT OF THE ELECTRICAL UNITS

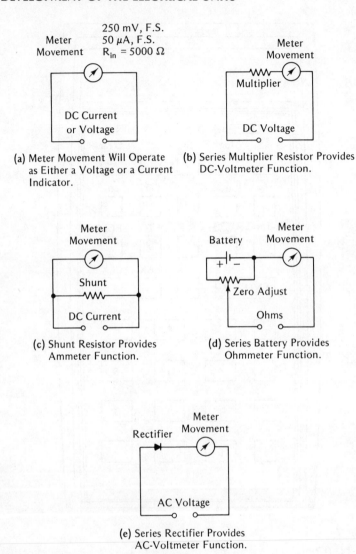

Figure 1-13. Skeleton multimeter instrument circuitry.

single-range voltmeter or as a single-range dc microammeter. In this case, the meter movement has a full-scale voltage value of 0.25 V (250 mV) and an internal resistance of 5,000 ohms. Because of this voltage-resistance relation, the full-scale current value of this meter movement is 50 μA. The movement has a sensitivity that is equal to the reciprocal of its full-scale current value, or 20,000 ohms-per-volt. A higher voltage range is provided by a multiplier resistor. A higher current range is provided by a shunt resistor while a series battery provides an ohmmeter range and a series rectifier provides an ac-voltage range.

1-11 Fundamental Test and Measurement Techniques

Most of the measurements made in electronic circuitry concern printed-circuit conductors, as shown in Figure 1-14. It is essential to avoid accidental short-circuits between adjacent conductors. A short-circuit for even a fraction of a second can cause catastrophic damage to semiconductor devices. Therefore, small test prods should be employed and should be applied with care. Printed-circuit conductors are generally coated with an insulating varnish or other substance. Good contact is required for reliable measurements. The operator should utilize sharply pointed test tips as illustrated in Figure 1-15. If a sharp point is permitted to "skate" across a printed-circuit conductor, it is possible to open-circuit the conductor. In the event that this accident should occur, the open-circuit can be repaired with a small drop of solder.

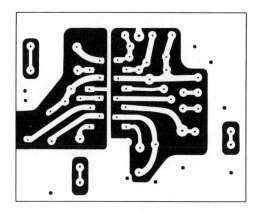

Figure 1-14. Example of printed-circuit board conductor layout.

All relations among measured electrical values are based on Ohm's law and upon Joule's law, as summarized in Figure 1-16. In other words, Ohm's law states that $I = E/R$, and Joule's law states that $W = EI$. By substitution, power may be expressed in terms of current and resistance, or of voltage and resistance. Current may be expressed in terms of power and voltage, or of power and resistance. Voltage may be expressed in terms of current and resistance, of power and current, or of power and resistance. Resistance may be expressed in terms of voltage and current, of voltage and power, or of current and power.

20 DEVELOPMENT OF THE ELECTRICAL UNITS

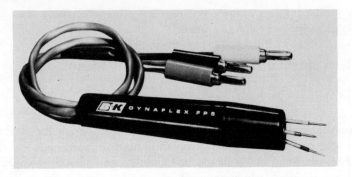

Figure 1-15. Typical test leads designed for use in printed circuitry. (*Courtesy*, B & K Precision, Div. of Dynascan Corp.)

Figure 1-16. Summary of relations among measured values of voltage, current, resistance, and power.

1-12 Instruments Utilized in Various Frequency Ranges

Basic measurement procedures are concerned with zero frequency (direct current), audio frequencies up to 20 kHz, ultrasonic frequencies up to hypersonic frequencies (five times the audio-frequency limit), video frequencies from zero to 4 MHz, intermediate frequencies from 455 kHz to 47.25 MHz, broadcast radio frequencies from 550 kHz to 108 MHz, broadcast television frequencies from 54 to 890 MHz, two-way radio frequencies from 37.02 to 465.5 MHz, and digital equipment frequencies up to 50 MHz. In each of these ranges, certain instruments are generally employed, as listed in Table 1-4. Some instruments, such as the transistor voltmeter (TVM) are utilized in all areas. Others, such as the data domain analyzer, are used only in the digital area. An oscilloscope that is designed

Table 1-4
Instruments Utilized in Various Frequency Ranges

Audio Frequencies: 20 Hz to 20 kHz	Video Frequencies: 0 to 4 MHz	Intermediate Frequencies: 455 kHz to 47.25 MHz	Broadcast Radio Frequencies: 550 kHz to 108 MHz
Audio Oscillator	Video-sweep Generator	AM Signal Generator	AM Signal Generator
Audio Sweep Generator	Test-pattern Generator	IF Sweep Generator	FM Signal Generator
Oscilloscope	Square-wave Generator	Test-pattern Generator	Signal Tracer
Harmonic Distortion Meter	Oscilloscope	Oscilloscope	Oscilloscope
Intermodulation Analyzer	Semiconductor Tester	Signal Tracer	Semiconductor tester
Square-wave Generator	Transistor Multimeter	Semiconductor Tester	Transistor Multimeter
Tone-burst Generator	Capacitor Checker	Transistor Multimeter	Capacitor Checker
Impedance Bridge	Color-bar Generator	Capacitor Checker	Stereo FM Multiplex Generator
Semiconductor Tester		Stereo FM multiplex Generator	Frequency Counter
Transistor Multimeter			
Capacitor Checker			

Broadcast Television Frequencies: 54 to 890 MHz	Two-way Radio Frequencies: 37.02 to 465.5 MHz	Digital Equipment Frequencies: Up to 50 MHz
VHF and UHF Sweep and Marker Generators	Lab-type AM and FM Signal Generators	Logic Probe
Test Pattern Generator	Deviation Meter	Logic Pulser
Oscilloscope	RF Wattmeter	Logic Clip
Semiconductor Tester	Digital Frequency Counter	Logic Comparator
Transistor Multimeter	Transistor Multimeter	Current Tracer
Capacitor Checker	Tube Tester	Oscilloscope
	Transistor Tester	Data Domain Analyzer
	SWR Meter	Transistor Multimeter

22 DEVELOPMENT OF THE ELECTRICAL UNITS

for audio-frequency measurements has considerably different characteristics from one that is designed for digital-pulse measurements. Similarly, a signal generator that is designed for broadcast radio measurements has different ranges and characteristics from one that is designed for two-way radio measurements.

Review Questions

1. Why is the electrometer of historical significance?
2. What was the function of the voltaic pile?
3. Name the established mechanical units of measurement.
4. Distinguish between absolute units and concrete standards.
5. How was the unit of electric current originally established?
6. Who produced the first standard cell?
7. When was the SI System adopted?
8. Why are logarithmic scales used on some analog instruments?
9. Define a true rms instrument.
10. Explain the term "overranging."
11. What is the function of a multiplier resistor?
12. Describe the function of a shunt resistor for a meter movement.
13. Why must an instrument rectifier be employed in an ac voltmeter?
14. If a meter movement has a full-scale current value of 50 μA, what is its sensitivity value?
15. Upon what two electrical laws are all relations among measured electrical values based?

chapter two

Statistics and Measurement Errors

2-1 General Considerations

The term *experiment* is quite broad, in that it may denote any measurement project from an afternoon's set of laboratory activities to an extended research investigation. Therefore, the following discussion is of a very general nature. All scientific or technological projects have many points in common, regardless of the scope of the undertaking. An essential component in all measurement procedures is a careful and inquiring mental attitude in approaching a problem. A thorough understanding of basic principles is necessary in order to visualize an approach (or preferably several approaches) to the problem. Imagination, ingenuity, and creativity can lead the worker to solutions beyond the range of ordinary routine processes. The ability to visualize possible sources of error and potential disturbing factors that may interfere with realization of accurate measurements is also needed. Otherwise, the outcome may range from second-class, to misleading, to actually erroneous. Measurements combine both art and science. A knowledge of scientific principles is necessary, but it is not sufficient—skill in perfecting techniques is essential for achieving high-quality results.

Measurements involve the use of instruments of one sort or another. In general, an instrument provides a physical means of making measurements of greater refinement than are possible to the unaided human faculties.

26 STATISTICS AND MEASUREMENT ERRORS

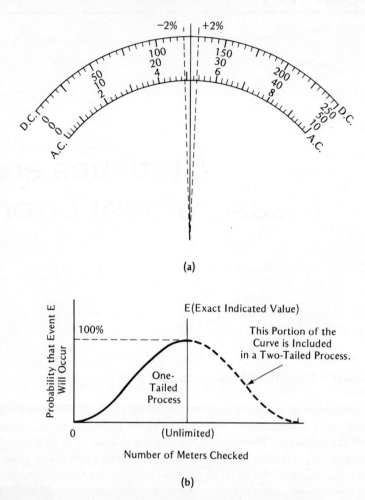

Figure 2-1. A qualitative example of the probability of an exact indication: (a) instrument accuracy rating; (b) normal curve of distribution.

Instruments also enable measurements that human faculties would be unable to sense or to measure. This consideration is particularly true in the field of electricity. Many ingenious and sophisticated instruments have been developed over the years. Our procedures and techniques in electrical measurement situations involve understanding and applying these instruments to obtain reliable and accurate results. Beginners sometimes suppose that if an instrument indicates a particular value, this value is absolutely correct. On the other hand, the experienced metrologist recognizes that any instrument whatsoever necessarily has some error of indication, although this error may be very small under optimum conditions.

2-2 PRECISION AND ACCURACY CONSIDERATIONS

As an illustration, Figure 2-1 shows the meaning of a ±2 percent of full scale accuracy rating on a voltmeter. In (a), the solid line depicts a scale indication of 125 volts. Since the full-scale value is 250 volts, and the accuracy rating of the meter in this example is ±2 percent, it follows that the scale indication could be as much as 5 volts high or 5 volts low. As shown by the dotted lines, the range of scale indication error in this example extends from 120 volts to 130 volts. Therefore, it would be incorrect to assume that the value of the voltage measured in this example is *exactly* 125 volts. The correct specification of this measurement is 125 volts ±5 volts, or 125 volts ±4 percent. Note that, in the case of a half-scale reading, the possible error in the indicated value is double the full-scale accuracy rating of the meter.

Distribution of Probable Error

Next, consider the probable distribution of error within the limits of possible error. A qualitative presentation of probable distribution is shown in Figure 2-1(b). If a large number of voltmeters were carefully checked for calibration at 125 volts, their scale-indication values would be expected to cluster around a 125-volt value as exhibited by the normal curve of distribution. It would be extremely unlikely that any one of the meters would indicate a value of 125 volts so accurately that additional and more refined calibration tests would confirm an assumption of absolutely exact indication. It also would be extremely unlikely that any one of the meters would indicate a value outside of the limits of possible error corresponding to the accuracy rating of the meter. Therefore, we may conclude:

1. It is more probable that the true value of the voltage under measurement is nearer to 125 volts than it is to 120 volts or to 130 volts.
2. It is highly improbable that the true value of the measured voltage is less than 120 volts or more than 130 volts. This, however, cannot be stated as a fact until the measurement has been satisfactorily cross-checked.

2-2 Precision and Accuracy Considerations

Although the terms *accuracy* and *precision* are often employed, little distinction is observed between the meanings of these words. The following brief definitions distinguish between the terms as applied to measurement procedures:

Precise—Sharply or clearly defined.
Accurate—Conforming to truth.

28 STATISTICS AND MEASUREMENT ERRORS

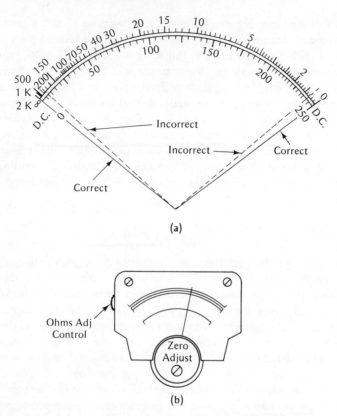

Figure 2-2. Examples of offset errors: (a) correct and incorrect zero adjustments for the voltage scale; correct and incorrect adjustments for the resistance scale; (b) example of zero-adjust and ohms-adjust controls.

As an illustration of the difference in meaning between these two terms, consider that we are using a volt-ohmmeter which is normal in every respect, except that its zero-set control is incorrectly adjusted (see Figure 2-2). In turn, the instrument is as *precise* as ever, inasmuch as we can make voltage "measurements" within its rated indication tolerance. These voltage readings will be consistent and "clearly defined." On the other hand, the voltage readings taken from this instrument are *not accurate*, inasmuch as they do not conform to truth. These same observations apply to resistance measurements that may be made with the ohms-adjust control incorrectly set, so that the pointer is offset from its correct zero-ohms position.

An example of a decade resistor box is shown in Figure 2-3. We may compare two decade resistor boxes that provide resistance increments of 1, 10, 100, and 1000 ohms per step. For example, one resistor box with economical construction may have a rated accuracy of ±1 percent. On the other hand, another resistor box with high-quality construction may have a

2-2 PRECISION AND ACCURACY CONSIDERATIONS 29

Figure 2-3. Example of a decade resistance box.

rated accuracy of ±0.1 percent. Since either box may be set to any value up to 10,000 ohms in 1-ohm steps, both of them have the same *precision*. Insofar as *accuracy* is concerned, however, the latter has ten times better accuracy than the former.

As another illustration of the distinction between accuracy and precision, consider two voltmeters of the same make and model. Assume that these meters have carefully ruled scales, knife-edge pointers, and mirror-backed scales to minimize parallax errors. It follows from previous discussion that both meters may be read to the same *precision*. With reference to the instrument circuitry exemplified by Figure 2-4, we will stipulate that the meters are operated on their 1200-V range. We will further stipulate that the 4.8-meg multiplier resistor in one of the meters has increased 30 percent in value. Of course, this fault results in a subnormal scale indication. Although both instruments may be read to the same *precision*, their *accuracies* will be quite different.

If a measured value is to be determined with an accuracy to a specified number of digits, it is necessary that the measuring equipment have a precision in this order. Otherwise stated, *precision is a necessary prerequisite to accuracy*, but as explained above, *precision does not guarantee accuracy*. Accuracy is achieved by careful measurement in terms of an accurately known standard. Precision is essential in the detection of possible inaccuracy, as in comparative measurements of a quantity by two methods. Precision alone, however, does not ensure accuracy. A pointer position on a scale can be read to higher precision if a mirror-backed scale is utilized, as shown in Figure

30 STATISTICS AND MEASUREMENT ERRORS

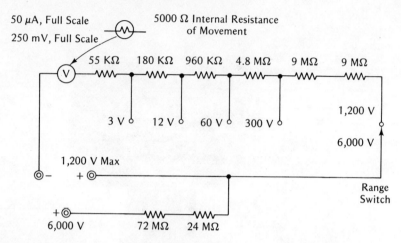

Figure 2-4. Example of series multiplier resistor circuitry in a dc voltmeter.

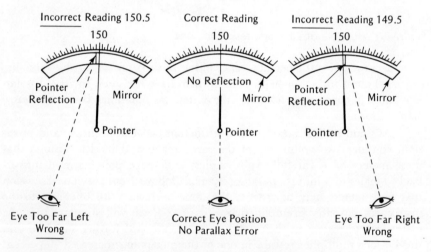

Figure 2-5. A mirror-backed scale serves to minimize parallax error. (*Courtesy*, B & K Precision, Div. of Dynascan Corp.)

2-5. In other words, a mirrored scale minimizes parallax error (a form of observational error). It is impossible to completely eliminate observational error in reading analog scales. Observational errors tend to follow the normal curve of distribution that was shown in Figure 2-1.

Precision Factors

A set of readings is said to show precision if the results agree among themselves. However, the set of readings may lack accuracy. Any systematic

2-2 PRECISION AND ACCURACY CONSIDERATIONS

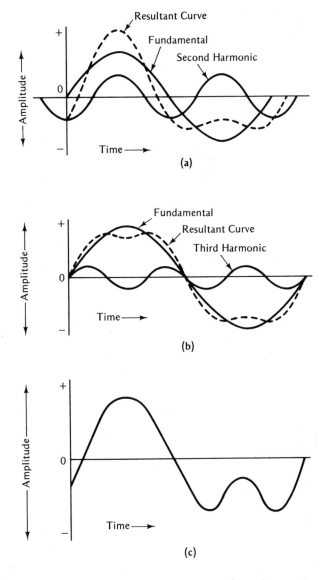

Figure 2-6. Basic second- and third-harmonic distortions: (a) Second-harmonic distortion; (b) third-harmonic distortion; (c) combined second- and third-harmonic distortions.

error that enters into the readings will impair their accuracy accordingly. Therefore, it is good practice to employ more than one method of measurement so that an effective cross-check is provided. It will be shown that systematic errors can arise from various sources; accordingly, cross-checks should be planned to avoid confusion from these possibilities. Thus, there is a possibility of shortcomings in the measuring instrument, of inadvertent misapplication of an instrument, of interfering environmental factors such as temperature or humidity, and of systematic observational errors.

If a multiplier resistor in a voltmeter is off-value, there is a shortcoming in the instrument. On the other hand, if a voltmeter indicates incorrectly because of a characteristic of the applied voltage, the instrument has been inadvertently misapplied. For example, an ac voltmeter may employ instrument rectifiers, or it may have thermocouple construction. A voltmeter with instrument rectifiers can indicate the true rms value of a sine-wave voltage, but it cannot indicate the true rms value of a sine-wave voltage with harmonics. A voltmeter with thermocouple construction, however, will indicate the true rms value of either a sinusoidal or of a nonsinusoidal voltage waveform. (See Figure 2-6.) As another illustration, a voltmeter has a certain *current burden*. Thus, a 1000 ohms-per-volt instrument draws 20 times as much current from the circuit under test as does a 20,000 ohms-per-volt instrument. If a voltmeter draws enough current to disturb the action of the circuit under test, the resulting scale indication will be incorrect—the voltmeter has been inadvertently misapplied.

2-3 Voltmeter Sensitivity

Since lack of accuracy in voltage measurement often results from unsuspected circuit loading, it is instructive to consider the essentials of voltmeter sensitivity. It follows from the configuration of Figure 2-4 that this dc voltmeter will have a different value of input resistance on each range. Thus, the instrument has an input resistance of 60 kilohms (kΩ) on its first range, and an input resistance of 110 megohms (MΩ) on its highest range. The *sensitivity* of the voltmeter is defined in terms of its ohms-per-volt rating. On each range, its input resistance on that range, divided by the associated full-scale indication, is a constant. In this example, the constant is 20,000. In turn, this voltmeter is described as a 20,000 ohms-per-volt instrument. The sensitivity of a dc voltmeter is also equal to the full-scale current demand of the meter movement.

As shown in Table 2-1, a 1000 ohms-per-volt meter has an input resistance of 2500 ohms on its 2.5-V range, whereas a 100,000 ohms-per-volt instrument has an input resistance of 250,000 ohms on its 2.5-V range. This is a 100-to-1 difference in loading action (current demand), and it can be a highly significant factor in measurement accuracy considerations. As an illustration, consider the measurements depicted in Figure 2-7. A potential of five volts is measured in a high-resistance circuit with three types of voltmeters. One voltmeter has a sensitivity of 1000 ohms-per-volt, the second voltmeter has a sensitivity of 20,000 ohms-per-volt, and the third voltmeter

Table 2-1

	DC Voltmeter Input Resistance		
Meter Sensitivity	1,000 Ω/V	20,000 Ω/V	100,000 Ω/V
2.5 V Range	2,500 Ω	50 KΩ	250 KΩ
10 V Range	10 KΩ	200 KΩ	1 Meg
50 V Range	50 KΩ	1 Meg	5 Meg
250 V Range	250 KΩ	5 Meg	25 Meg
1,000 V Range	1 Meg	20 Meg	100 Meg
5,000 V Range	5 Meg	100 Meg	500 Meg

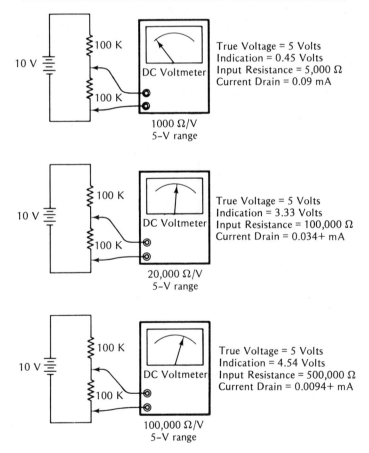

Figure 2-7. Examples of circuit-loading errors in voltage measurement.

34 STATISTICS AND MEASUREMENT ERRORS

has a sensitivity of 100,000 ohms-per-volt. Observe that the low-sensitivity voltmeter indicates only 0.45 V, whereas the true potential is five volts. The medium-sensitivity voltmeter indicates 3.33 V, and the high-sensitivity voltmeter indicates 4.54 V. If a solid-state multimeter were utilized that has an input resistance of 10 MΩ, it would indicate 4.97 + V, and the error would be approximately 0.6 percent.

Current-measurement Errors

Indication errors also may be encountered owing to misapplication of current-measuring instruments. In other words, a current-measuring instrument has a finite value of internal resistance, and this resistance is greatest on the lowest range of the instrument. The internal resistance of a current-measuring instrument disturbs the action of low-resistance circuits more than high-resistance circuits. A practical example of circuit disturbance is shown in Figure 2-8. Here, a 50 μA meter with an internal resistance of 5000 ohms is connected in series with the base circuit of a transistor. The current value indicated by the meter will be considerably subnormal, because the instrument adds 5000 ohms of resistance in series with the base circuit. A microammeter with a substantially lower value of internal resistance must be employed in this situation.

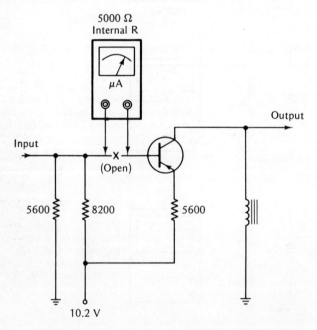

Figure 2-8. Example of circuit-disturbance error in current measurement.

2-3 VOLTMETER SENSITIVITY 35

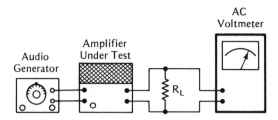

Figure 2-9. Frequency response of the ac voltmeter must be at least equal to that of the amplifier under test.

Another type of meter misapplication that can cause incorrect voltage indication is shown in Figure 2-9. Here, an ac voltmeter is utilized to check the frequency response of an amplifier. In turn, the frequency response of the voltmeter must be at least equal to that of the amplifier under test. Some audio amplifiers have both extended low-frequency response and extended high-frequency response. Accordingly, the voltmeter must have full response at the lowest and at the highest frequencies of test. Otherwise, deficiencies in meter response would be falsely attributed to the amplifier under test.

Still another illustration of voltmeter misapplication is seen in Figure 2-10. The base voltage of a UHF oscillator transistor is being measured. Observe that the input cable to the voltmeter is connected directly to the base terminal (an isolating resistor is not employed). Consequently, the input cable operates as a tuned stub with respect to UHF oscillator action. This tuned-circuit disturbance can shift the oscillating frequency, change the mode of operation, or "kill" the oscillator action, depending upon the length of the input cable. In turn, the normal dc base voltage may be modified substantially. To avoid, or at least to minimize, this possible source of incorrect indication, the dc voltmeter should be provided with an isolating resistor in the tip of its input cable. Service-type, solid-state voltmeters are often designed with a 0.5 or 1 meg dc probe for this reason.

Distinction should be made between *current leads* and *voltage leads* for electrical measuring instruments. In other words, instrument leads have more or less resistance. When a high-impedance voltmeter is applied, it draws a very small amount of current from the circuit under test. Consequently, the voltage indication is not affected (IR lead drop is negligible) although small-diameter lead conductors may be utilized. On the other hand, when a high-current ammeter is applied, the high-valued current in the circuit under test also flows through the ammeter (and its test leads). Consequently, unless large-diameter lead conductors are employed, the IR drop across the test leads will reduce the current value and produce an incorrect scale indication.

36 STATISTICS AND MEASUREMENT ERRORS

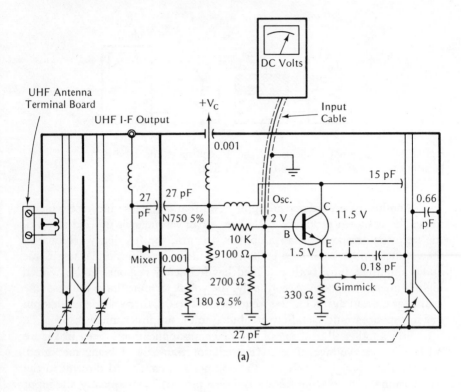

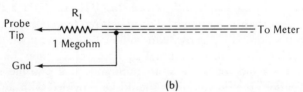

Figure 2-10. Input cable capacitance can disturb oscillator action: (a) Example of direct cable applied in a UHF oscillator circuit; (b) typical isolating probe for a DC voltmeter.

2-4 Error Computation

It is sometimes feasible to compute a correct scale indication when the conditions responsible for an incorrect scale indication are known. As a basic illustration, consider measurement of a square-wave voltage with an ac voltmeter that contains a half-wave instrument rectifier. In Figure 2-11, the meter scale is calibrated in rms values. When a half-wave instrument rectifier is provided, the instrument current consists of half-sine waves. The average value of this waveform is 0.318 of its peak value. Although the meter

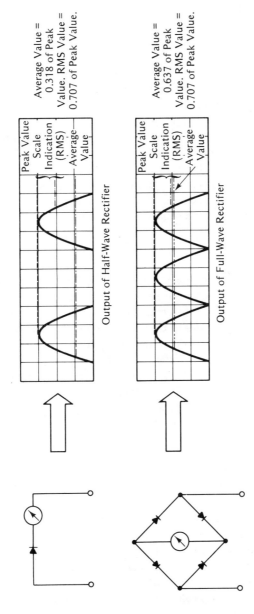

Figure 2-11. Meter movement responds to average value of rectified waveform; scale indicates rms value of input sine waveform.

38 STATISTICS AND MEASUREMENT ERRORS

movement responds to the average value of the current waveform, the instrument scale is calibrated in rms values of a sine wave. The rms value of a sine wave is equal to 0.707 of peak. Therefore, if 1 volt dc is applied to an ac voltmeter that has a half-wave rectifier, the scale indication will be 2.22 V. Consider next how these factors determine the instrument response to a square-wave input.

Referencing Figure 2-11, observe that a square-wave voltage with a peak value of A will produce a scale indication of 1.11 A on an ac voltmeter that employs a half-wave instrument rectifier. This is necessarily the case, because the scale indication would be 2.22 A for an applied dc voltage with a value of A. Inasmuch as the square wave is on for half the time and off for half the time with respect to a half-wave instrument rectifier, the scale indication must be one-half of 2.22 A, or 1.11 A. In other words, the operator can measure the peak amplitude of a square-wave voltage in this situation by computation, reducing the scale indication by 9.91 percent to calculate the peak value of the square-wave voltage.

Response to Full-wave Instrument Rectifier

Next, consider the response of an ac voltmeter with a full-wave instrument rectifier to a square-wave voltage. We observe in Figure 2-11 that the average value of a full-rectified sine waveform is equal 0.637 of its peak value. As noted previously, the rms value of a sine wave is equal to 0.707 of peak. Therefore, if a dc voltage is applied to an ac voltmeter that has a full-wave instrument rectifier, the scale indication will be equal to 0.707/0.637, or to 1.11 times the value of the applied dc voltage.

Since an applied square-wave voltage has both positive and negative excursions (see Figure 2-12), each with an amplitude of A, it is evident that the voltmeter will respond in the same manner as if a dc voltage with an amplitude of A were applied. Therefore, the scale indication will be equal to 1.11 times the peak amplitude of the applied square-wave voltage. Accordingly, the operator can measure the peak amplitude of a square-wave voltage in this situation by calculation and reducing the scale indication by 9.91 percent.

Note in passing that some ac voltmeters utilize instrument circuitry wherein the meter movement responds to the peak value of the applied waveform. Since the majority of such instruments are designed for sine-wave applications, the instrument scale is generally calibrated in terms of rms values for a sine wave. Consequently, if the amplitude of a square-wave voltage is measured with a peak-responding ac voltmeter, the operator must determine its peak value by calculating 70.7 percent of the scale indication. Since the rms value of a dc voltage is equal to its magnitude, the rms value of a square wave is evidently equal to its peak amplitude, as indicated in Figure 2-13. Since scale-correction calculations and rms-value computations

2-4 ERROR COMPUTATION 39

Note: "A" is the Peak Amplitude of the Waveform.

Waveform	Meter Response	Scale Reading
Square Wave	Peak	0.707 A
	$\frac{1}{2}$ Wave Average	1.11 A
	Full-Wave Average	1.11 A
Sawtooth Wave	Peak	0.707 A
	$\frac{1}{2}$ Wave Average	0.555 A
	Full-Wave Average	0.555 A
$\frac{1}{2}$ Rectified Sine Wave	+Peak	0.707 A
	$+\frac{1}{2}$ Wave Average	0.707 A
	Full-Wave Average	0.354 A
Full Rectified Sine Wave	Peak	0.707 A
	$+\frac{1}{2}$ Wave Average	1.414 A
	Full-Wave Average	0.707 A

Figure 2-12. Response of basic ac voltmeters to common complex waveforms.

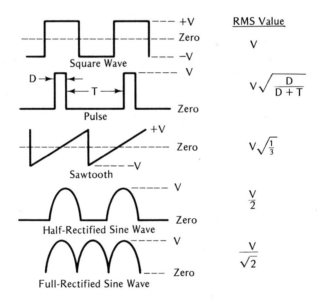

Waveform	RMS Value
Square Wave	V
Pulse	$V\sqrt{\frac{D}{D+T}}$
Sawtooth	$V\sqrt{\frac{1}{3}}$
Half-Rectified Sine Wave	$\frac{V}{2}$
Full-Rectified Sine Wave	$\frac{V}{\sqrt{2}}$

Figure 2-13. Rms values of some common complex waveforms.

40 STATISTICS AND MEASUREMENT ERRORS

require the use of higher mathematics for various waveforms, most instrument operators prefer to consult charts as illustrated in Figures 2-12 and 2-13 when a complex waveform is encountered.

2-5 Ohmmeter Scale Resolution

Most analog ohmmeters have nonlinear scales. In turn, the effective resolution changes from one end of the scale to the other. As shown in Figure 2-14, the left-hand end of the resistance scale provides comparatively poor resolution, whereas the right-hand end of the scale provides comparatively good resolution. Therefore, it is good practice to operate an ohmmeter on a range that provides pointer deflection above the half-scale point. In other words, if a resistor with a value of approximately 200 ohms is to be measured, the operator can obtain a much more accurate measurement by operating the ohmmeter on its $R \times 100$ range, or on its $R \times 10$ range, instead of its $R \times 1$ range. Because an ohmmeter scale is cramped at its high-resistance end, the absolute error usually increases as the pointer indicates higher values.

The rated accuracy of an ohmmeter is commonly stated in degrees of arc. This rating refers basically to the accuracy of the dc-voltage function of the instrument. As an illustration, the dc-voltage accuracy might be rated at ± 2 percent of full scale. On the 10-V scale, this rating indicates an accuracy of ±0.2 V. Correspondingly, a certain number of degrees of arc defines this interval of 0.4 V as depicted in Figure 2-15. This arc also defines the accuracy of the ohmmeter scale. Note, however, that this accuracy will not be realized if the ohmmeter is operated with a weak battery.

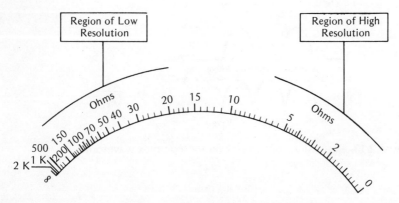

Figure 2-14. Resolution becomes progressively poorer toward the left-hand end of the scale.

2-5 OHMMETER SCALE RESOLUTION 41

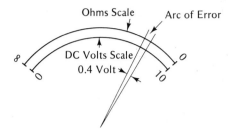

Figure 2-15. An accuracy rating of ±2 percent of full scale corresponds to an interval of 0.4 volt on a 10-volt scale.

42 STATISTICS AND MEASUREMENT ERRORS

Review Questions

1. What is the role of one's mental attitude in an approach to a measurement problem?

2. Why do measurement techniques involve both art and science?

3. Explain an instrument accuracy rating of ± 2 percent of full-scale indication.

4. Distinguish between the limits of probable error and the probable distribution of error.

5. Define the term *precision*; define the term *accuracy*.

6. Describe the term "waveform error".

7. If a square wave has a peak value of 1 volt, what is its rms value?

8. If a half-rectified sine wave has a peak value of 1 volt, what is its rms value?

9. If a full rectified sine wave has a peak value of 1 volt, what is its rms value?

10. How is the arc of error derived for an ohmmeter scale?

11. An ac voltmeter employs a half-wave instrument rectifier, and its scale is calibrated to indicate rms values for a sine-wave input. If a square-wave input is applied to the voltmeter, does the pointer indicate its rms value? Why?

12. An ac voltmeter employs a full-wave instrument rectifier, and its scale is calibrated to indicate rms values for a sine-wave input. If a square-wave input is applied to the voltmeter, does the pointer indicate its rms value? Why?

13. A half-rectified sine-wave input is applied to the ac voltmeter in problem 11. Does the pointer indicate its rms value? Why?

14. The half-rectified sine-wave input is applied to the ac voltmeter in problem 12. Does the pointer indicate its rms value? Why?

15. A full-rectified sine-wave input is applied to the ac voltmeter in problem 12. Does the pointer indicate its rms value? Why?

chapter three

Resistance Measurements

3-1 General Considerations

Various kinds of resistance are measured in electrical and electronics technology and suitable types of instruments must be utilized. Note the basic classifications of resistance that must be considered:

1. *Linear Resistance.* A linear resistance maintains the same E/I ratio at any voltage or current value within its rated limits.
2. *Nonlinear Resistance.* A nonlinear resistance exhibits a changing E/I ratio at different voltage or current values within its rated limits.
3. *Positive Resistance.* A positive resistance draws more current as the applied voltage is increased within its rated limits.
4. *Negative Resistance.* A negative resistance draws less current as the applied voltage is increased within its rated limits. Conversely, a negative resistance draws more current as the applied voltage is decreased within its rated limits.
5. *DC Resistance.* The dc resistance of a wire, for example, is defined as its E/I ratio in response to an applied dc test voltage within its rated limits.
6. *AC Resistance.* The ac resistance of a wire, for example, is defined as its E/I ratio in response to an applied ac test voltage of stipulated

43

44 RESISTANCE MEASUREMENTS

frequency within its rated limits. In general, the ac resistance value will be greater than the dc resistance value.

7. *Incremental Resistance* (also termed *dynamic resistance*). The ac resistance measured over a relatively small interval of a nonlinear resistance characteristic.
8. *Tangible Resistor.* A physical object that exhibits electrical resistance, such as a wire-wound resistor.
9. *Intangible Resistance.* The resistance characteristic of "empty" space within a device, such as the plate-cathode resistance of a vacuum diode, measured under specified operating conditions.
10. *Light-Dependent Resistor* (LED). A light-dependent resistor with constant applied voltage which changes its E/I ratio in response to the intensity of incident light.
11. *Thermal Resistor* (thermistor). A thermistor with constant applied voltage which changes its E/I ratio in response to a variation in ambient temperature.

In addition to the foregoing basic classifications of resistance, we will consider various subclassifications. For example, metal wire exhibits a fixed E/I ratio at any voltage or current value within its rated limits under standard conditions. On the other hand, if the wire is mechanically strained, its E/I ratio changes in response to the applied force. This is the basis of the *strain gage*. Again, there are three basic subclassifications of temperature coefficients of resistance. Thus, a copper wire has a positive temperature coefficient of resistance. This means that its E/I ratio increases as the ambient temperature increases. On the other hand, a thermistor has a negative temperature coefficient of resistance; its E/I ratio decreases as the ambient temperature increases. Specialized resistors are fabricated to have a zero temperature coefficient of resistance. Their E/I ratio remains constant over a rated range of ambient temperature change.

3-2 Semiconductor Junction Resistance

Semiconductor diodes and transistors have junctions that are essentially rectifiers. In other words, the junction conducts for one polarity of applied voltage, but does not conduct for the other polarity of applied voltage. In practice, a junction may exhibit more or less leakage resistance. This leakage permits some value of reverse current flow. The ratio of forward resistance to reverse resistance is called the *front-to-back ratio* of the junction. Junction resistance is ordinarily measured with an ohmmeter, such as illustrated in Figure 3-1. Front-to-back ratios are measured with the hi-pwr ohms function of the multimeter. Consider next the technical distinction between a lo-pwr ohms function and a hi-pwr ohms function.

3-2 SEMICONDUCTOR JUNCTION RESISTANCE 45

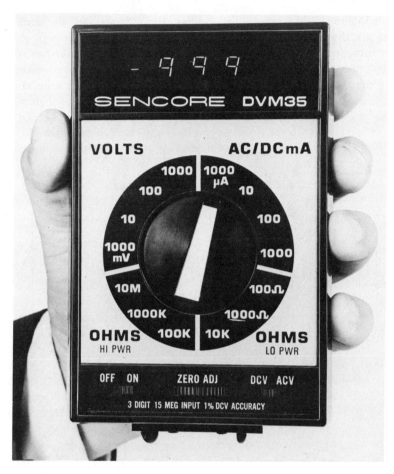

Figure 3-1. A digital multimeter with hi-pwr and lo-pwr ohmmeter functions. (*Courtesy,* Sencore)

A hi-pwr ohms function provides conventional ohmmeter tests. The source voltage employed is typically 1.5 V; a lo-pwr ohms function utilizes a source voltage of 0.08 V. This value of test voltage is below the threshold level of semiconductor junction conduction. Therefore, when the lo-pwr ohms function is used, all normal semiconductor junctions appear to be open circuits, regardless of the test-voltage polarity. This is a useful feature in troubleshooting solid-state circuitry, for example, because the values of many resistors can be measured in-circuit. Stated otherwise, a test voltage of 0.08 V will not "turn on" a normal semiconductor junction. This test voltage, however, is as useful as a higher test voltage for measuring the resistance value of a composition resistor, for example.

46 RESISTANCE MEASUREMENTS

In Figure 3-2, the exemplified configuration "looks like" the semiconductors are disconnected when a lo-pwr ohmmeter is used, as depicted in (b). In turn, it is practical to measure the value of each resistor in-circuit, provided that the lo-pwr function of the ohmmeter is employed. Note in passing that it is assumed that all semiconductor junctions are normal and that all capacitors have negligible leakage resistance. In the event that any resistor happens to measure an abnormally low value when checked in-

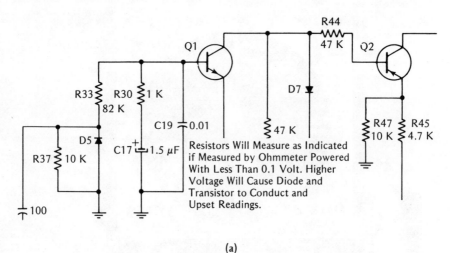

(a)

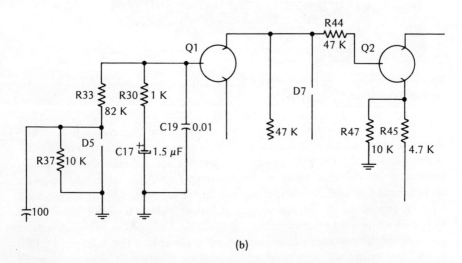

(b)

Figure 3-2. Resistance measurement in solid-state circuitry: (a) device junctions will conduct when forward-biased by test voltage from conventional ohmmeter; (b) how the circuit "looks" to a low-power ohmmeter.

circuit, it should be disconnected and its resistance value should be measured out-of-circuit. If its value then measures within rated tolerance, it is evident that the false in-circuit reading resulted from leakage resistance either in a semiconductor junction, or in a capacitor. However, if its value measures the same abnormally low value when checked out-of-circuit, it is obvious that the resistor is defective and must be replaced.

The basis for lo-pwr ohmmeter operation is seen in Figure 3-3. Observe

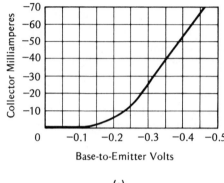

(a)

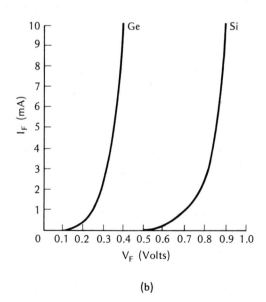

(b)

Figure 3-3. Transfer characteristic for a small-signal germanium transistor: (a) bias voltage versus collector current for a germanium transistor; (b) comparison of bias voltages for silicon and germanium transistors.

48 RESISTANCE MEASUREMENTS

that a germanium diode junction does not start to conduct until the forward bias voltage exceeds 0.1 V. A silicon diode junction does not start to conduct until the forward bias voltage exceeds 0.6 V. Note also that a germanium transistor does not pass collector current until its base-emitter forward bias voltage exceeds 0.1 V. It might be supposed that in-circuit resistance measurements could be made with a conventional ohmmeter, provided that the operator applied the test leads in a polarity such that semiconductor junctions in the path of current flow were reverse-biased by the test voltage of the ohmmeter. Although it is somewhat time-consuming, *this procedure is possible in various situations*. In other words, the operator must analyze the circuit under test to determine the required test-lead polarity. Again, this procedure is inapplicable in circuitry wherein the test-current flow can pass

Table 3-1
Resistance Chart for a UHF Tuner, VHF Tuner, and Panel Board*

Item	E	B	C
UHF	Tuner:		
Q1	3100 Ω	3300 Ω	0 Ω
Q2	1000 Ω	3000 Ω	10 K
Q3	1200 Ω	3200 Ω	1500 Ω

Item	E	B	C
VHF	Tuner:		
Q1	330 Ω	3500 Ω	2000 Ω
Q2	680 Ω	1500 Ω	1800 Ω
Q3	1000 Ω	4800 Ω	270 Ω

Item	E	B	C
Panel A Board			
Q4A	560 Ω	2200 Ω	3000 Ω
Q5A	2700 Ω^S	47 K ^{G}	2600 Ω^D
Q6A	470 Ω	1500 Ω	2200 Ω
Q7A	3800 Ω	13 K	2400 Ω

*Measurements taken with an ohmmeter that applies less than 0.08 V maximum between probe tips.

3-3 MEASUREMENT OF NONLINEAR RESISTANCE VALUES 49

through a forward junction, regardless of the test-voltage polarity. Therefore, a lo-pwr ohmmeter is the practical type of instrument for in-circuit resistance measurements. An example of a resistance chart is given in Table 3-1.

3-3 Measurement of Nonlinear Resistance Values

Semiconductor junction resistance is nonlinear. In turn, the resistance value that is indicated by an ohmmeter depends upon the test voltage that is applied, as illustrated in Figure 3-4. Note that if the applied test voltage is 0.4 V, the E/I ratio is 80 ohms. On the other hand, if the applied test voltage is 0.6 V, the E/I ratio is 40 ohms. Accordingly, the forward resistance value of a semiconductor junction will measure differently when checked with different ohmmeters. The junction resistance will also measure differently on different ranges of the same ohmmeter. Of course, if the

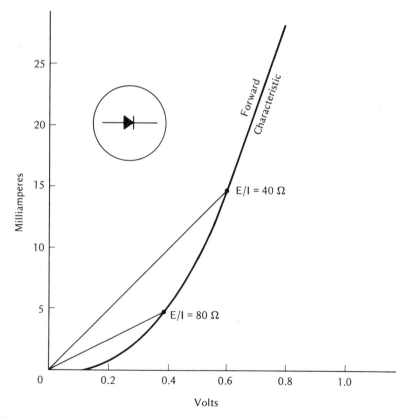

Figure 3-4. Resistance values at two points on a nonlinear characteristic.

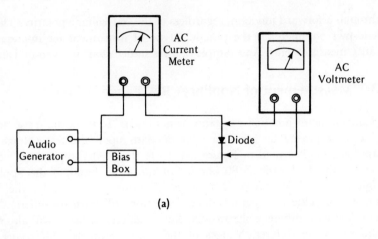

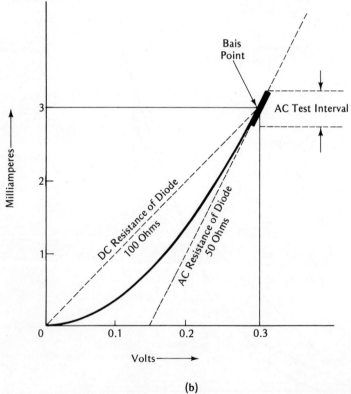

Figure 3-5. Comparison of ac and dc resistance values in a nonlinear system: (a) measurement of ac resistance; (b) ac and dc resistance values of a semiconductor diode.

junction resistance is checked with a lo-pwr ohmmeter, an infinite resistance will be indicated. Note in passing that *a semiconductor diode can be damaged by an old-style ohmmeter that applies excessive test voltage.*

Consider next the comparative dc and ac resistance values for a semiconductor diode, as depicted in Figure 3-5. The dc resistance at the bias point of 3 mA and 0.3 V is evidently 100 ohms—an E/I ratio of 0.3/0.003. Observe that the ac resistance (dynamic or incremental resistance) is defined by the E/I ratio for a small swing above and below the bias point. *The ac resistance is equal to the slope of the characteristic at the bias point.* In this example, the slope defines an E/I ratio of 50 ohms. Thus, the ac resistance of the diode is twice as great as its dc resistance at this particular bias point. In general, values of dc or ac resistance for a semiconductor junction are not meaningful unless the bias point is specified.

3-4 Positive Resistance and Negative Resistance

Positive resistance is characterized by an increase in current flow as the applied voltage is increased. On the other hand, negative resistance is characterized by a decrease in current flow as the applied voltage is increased. This distinction is seen in the E/I characteristic for a tunnel diode as shown in Figure 3-6. Observe that this characteristic has both positive and negative resistance intervals. The interval from the peak-current point to the valley-current point is a negative-resistance excursion,

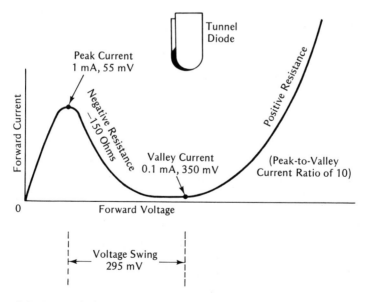

Figure 3-6. Forward characteristic for a tunnel diode.

while the interval from the origin to the peak-current point is a positive-resistance excursion. Also, the interval from the valley-current point through increasingly great current values is a positive-resistance excursion.

Negative resistance is generally regarded as an unstable circuit parameter. If the circuit has appreciable positive resistance, it becomes impossible to bias a tunnel diode into its negative-resistance region. In other words, the device operates as a switching circuit. As the applied voltage is increased past the peak-current point, the drop across the diode suddenly jumps to a value corresponding to the valley-current point. Then, if the applied voltage is decreased to a value below the valley-current point, the drop across the diode suddenly decreases to a value corresponding to the peak-current point. However, it should not be supposed that a tunnel diode always operates as a switching device. When connected into a suitable test circuit, a tunnel diode can be biased to any point along its negative-resistance interval.

The voltage across and the current through a tunnel diode can be measured in the same manner as for any type of semiconductor diode, provided only that the test circuit places a very low positive resistance in series with the tunnel diode. When this series positive resistance is less than the tunnel diode's negative resistance (at any point along its negative-resistance interval), the test circuit will be stable and the bias voltage can be adjusted to any desired value. The tunnel diode will not switch, under this condition, as the bias voltage is varied through the negative-resistance region of the device. Note that other devices and arrangements also can exhibit negative resistance. The tunnel diode, however, is the most widely used negative-resistance device. Note that when a tunnel diode is biased to a point along its negative-resistance characteristic it will operate as an amplifier.

3-5 Thermistor Resistance Measurements

Thermistors provide another instructive example of semiconductor resistance variation. A thermistor has a large, nonlinear, and negative temperature coefficient of resistance. Thermistors, such as depicted in Figure 3-7, are basically beads of certain metallic oxides. Typically, as the temperature of a thermistor is increased from 0 to 100°C, for example, the device resistance decreases from 1000 to 100 ohms. If the temperature of the thermistor is further increased to the point of burnout, its resistance decreases typically to a value of 10 ohms. Note that the resistance of a thermistor reads the same, regardless of ohmmeter polarity. Manufacturers generally specify thermistor resistance at standard room temperature.

Note that ballast resistors have a temperature coefficient of resistance that is the opposite of a thermistor, i.e., a ballast resistor has a positive coefficient. Thus, the resistance of an iron-wire/hydrogen ballast resistor increases rapidly as its temperature increases. Manufacturers usually specify

3-6 LIGHT DEPENDENT RESISTOR (LDR) RESISTANCE MEASUREMENT

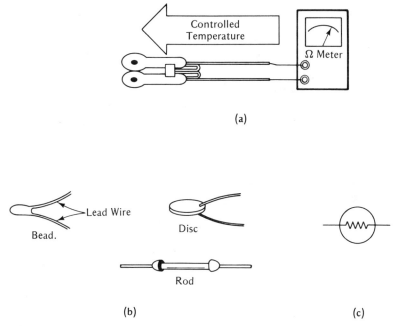

Figure 3-7. Thermistors: (a) ohmmeter check of thermistor resistance: (b) alternate types of thermistors; (c) symbol.

ballast resistance at standard room temperature. Note also that Globar resistors used in some radio receivers are sometimes called (incorrectly) "ballast resistors." A Globar resistor is not a ballast resistor in the technical sense of the term, because it has a negative temperature coefficient of resistance—the resistance of a Globar resistor decreases as its temperature increases.

3-6 Light Dependent Resistor (LDR) Resistance Measurement

Light-dependent resistors are semiconductor devices that decrease in resistance when exposed to light. The specified resistance value for an LDR is usually referenced to average room illumination. A typical LDR has a resistance that varies greatly with changes in ambient illumination. As an illustration, an LDR may vary in resistance from 100 ohms in bright light to 0.5 MΩ in complete darkness. With reference to Figure 3-8, the measured resistance value is the same, regardless of the test-voltage polarity. In other words, a CdS LDR is not a junction device. On the other hand, some LDR's have junction construction, and their measured resistance is polarity-dependent.

54 RESISTANCE MEASUREMENTS

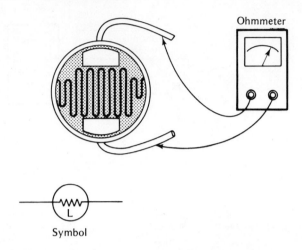

Figure 3-8. Resistance measurement of a CdS light-dependent resistor.

3-7 Internal, Input, and Output Resistance Measurements

To measure the output (internal) resistance of an instrument such as an audio generator, the test arrangement shown in Figure 3-9 may be used. First, the open-circuit voltage of the generator is measured—the operating frequency is inconsequential. Next, a variable resistor is connected across the generator output terminals, and its value is adjusted until the voltmeter indication is reduced to one-half of its open-circuit value. Then, the value of R is equal to the output resistance of the generator.

Consider next the measurement of input resistance to an amplifier, such as an audio amplifier. In Figure 3-10, the amplifier is driven by an audio generator. If a test frequency of 1 kHz is used, the amplifier's input resistance will have negligible reactive components and an accurate resistance measurement can be made. The amplifier is loaded by a power resistor with rated value and the voltage across the load is measured with an ac voltmeter. This output voltage value is noted. Next, a variable resistor is connected in series with the input lead to the amplifier and adjusted to reduce the output voltage to one-half of its original value. Then, the input resistance to the amplifier is calculated as follows:

$$R_{in} = R - R_s$$

where R_{in} is the amplifier input resistance

R is the input series resistance

R_s is the source (internal) resistance of the generator

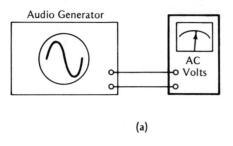

(a)

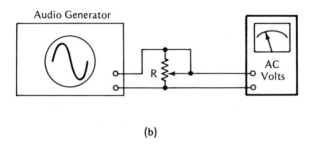

(b)

Figure 3-9. Generator output (internal) resistance: (a) output voltage is measured on open-circuit; (b) R is equal to generator output resistance when output voltage is reduced to one-half.

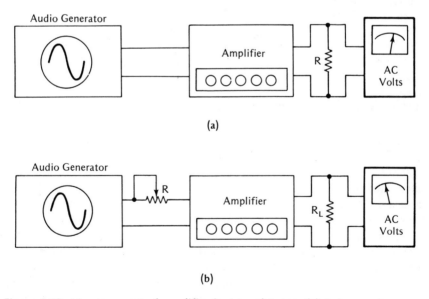

Figure 3-10. Measurement of amplifier input resistance: (a) output voltage is noted; (b) series resistance is inserted and adjusted to reduce output voltage to one-half (see text for calculation of R_{in}).

56 RESISTANCE MEASUREMENTS

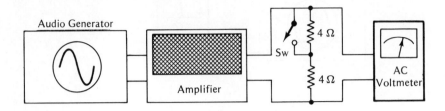

Figure 3-11. Measurement of amplifier output resistance.

Next, consider the measurement of the output resistance of an amplifier. (Refer to Figure 3-11.) Virtually all audio amplifiers can work into a 4-ohm or an 8-ohm load. Therefore, these two values have been chosen to illustrate the method of measurement. The amplifier is driven by an audio generator. If a test frequency of 1 kHz is used, the output resistance will be practically free from reactive components. One of the load resistors is provided with a switch, so that it can be connected or disconnected from the load circuit. An ac voltmeter measures the voltage developed across the load. First, the load voltage is measured for a load value of 8 ohms. Let this value be indicated as $V1$. Next, the load voltage is measured for a load value of 4 ohms, indicated as $V2$. If the amplifier output resistance is denoted by R_0, its value is related to $V1$ and $V2$ as follows:

$$R_0 = \frac{8(V2 - V1)}{V1 - 2V2}$$

3-8 Resistance Measurement with Megohmmeter

Special-purpose ohmmeters, called megohmmeters or meggers, operate at comparatively high voltage. A common type of megohmmeter operates at 600 volts; another type operates at 1000 volts. A typical megohmmeter is illustrated in Figure 3-12. This instrument is used to check for leakage resistance in television horizontal-output transformers and for insulation leakage in electric lines and cables. Its utility resides in its ability to detect insulation faults that are not apparent at a low test voltage, such as 1 volt. For example, if a high-voltage breakdown occurs in a horizontal-output transformer, the insulation may become charred between windings or from a winding to core. At a low test voltage, the char path may appear to have extremely high resistance, whereas a test at 600 or 1000 V causes arcing between adjacent char particles.

Figure 3-12. A 600-volt megohmmeter.

3-9 Resistance Strain Gage

A bonded type of resistance strain gage is depicted in Figure 3-13. It consists of a length of fine wire approximately 0.001 inch in diameter bent back and forth and cemented to a thin sheet of tissue paper. This paper is used to keep the wire strands in position, to insulate the wires from the surface under test, and to support the heavier connecting leads. To apply the strain gage, it is attached by a film of cement to the surface of the structural member to be tested, so that the fine wire is subjected to the same strains as the surface of the member. When the wire is stretched, its resistance increases. Many strain

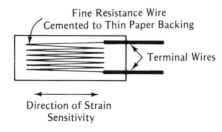

Figure 3-13. Plan of a simple strain gage

gages are made of constantan wire. A resistance of 120 ohms is typical. Because only small changes in resistance occur, a Wheatstone bridge is generally used to measure the resistance values under the applied strain forces.

3-10 Wheatstone Bridge Resistance Measurements

Highly accurate resistance measurements can be made with a Wheatstone bridge. In Figure 3-14, measurement is made by a null balance method. Since the null is unaffected by drift in battery voltage, and because a meter can be most accurately calibrated for zero current flow, the bridge method eliminates the major sources of inaccuracy inherent in the ohmmeter method. In the diagram, R_x denotes the *unknown* resistance value. If the value of R_2 is equal to that of R_1, then the value of R_3 is equal to that of R_x when the galvanometer is nulled. Thus, R_3 may be chosen as a precision potentiometer with an accuracy of 0.1 percent, and the setting of R_3 then indicates the value of R_x to a high degree of accuracy.

In practice, galvanometer protection must be provided, because sub-

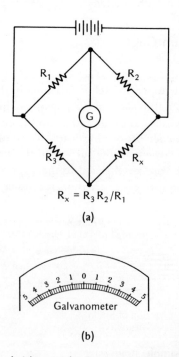

Figure 3-14. Wheatstone bridge: (a) basic circuit; (b) galvanometer scale.

stantial unbalance of the bridge would otherwise burn out the galvanometer. Older designs of Wheatstone bridges provide a galvanometer shunt with a push switch. The shunt greatly reduces the sensitivity of the galvanometer by connecting it in parallel with a comparatively low value of resistance. As the null is approached, the operator presses the push switch and thereby disconnects the shunt. In turn, the galvanometer operates at maximum sensitivity and permits an accurate null adjustment. More recent designs of Wheatstone bridges use semiconductor circuitry in combination with the null indicator, in order to provide automatic overload protection.

3-11 Measurement of Battery Internal Resistance

Measurement of the internal (source) resistance of a battery is depicted in Figure 3-15. Since the value of internal resistance depends upon the load current, a suitable value must be chosen for R. With the switch open, the operator notes the voltmeter reading. This is the open-circuit voltage E_{oc}, which is virtually equal to the EMF of the battery. Next, the switch is closed. The current meter indicates a load-current value I and the voltmeter indicates the terminal voltage of the battery under load E_{cc}. To calculate the value of R_{in}, the difference between E_{oc} and E_{cc} is divided by the measured current value. This method of internal-resistance measurement can be employed with any type of battery, power supply, or electric generator. For example, the internal resistance of a photovoltaic cell is measured as shown in Figure 3-16. The measured value at one intensity of incident light will not necessarily be the same as that at another intensity.

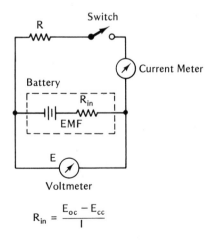

Figure 3-15. Measurement of battery internal resistance.

60 RESISTANCE MEASUREMENTS

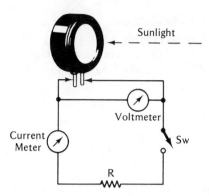

Figure 3-16. Measurement of the internal resistance of a photo-voltaic cell.

3-12 Typical Junction Resistance Values

Although it is not feasible to state "normal" ranges of junction-resistance values for bipolar and unipolar transistors indicated by various types of ohmmeters, it is helpful to note the general trend of these measured values. (Refer to Figure 3-17.) Indicated values were measured with a 20,000 ohms-per-volt meter with a 1.5-volt ohmmeter battery. As a technician gains experience with a particular ohmmeter in checks of normal and defective transistors (and diodes), guidelines are established for "weeding out" defective junction-type devices.

3-13 In-Circuit Measurement Expedient

In various situations, in-circuit resistance measurements may not appear to be feasible, owing to shunt arrangements. As an illustration, if a diode is connected across a resistor, a meaningful front-to-back resistance ratio check cannot be made unless the diode is disconnected from its circuit. However, it is seldom necessary to unsolder a diode or other device in order to make resistance measurements. As shown in Figure 3-18, it is usually possible to disconnect one end of the device from its circuit by making a slit with a razor blade across a suitable printed-circuit conductor. In turn, one end of the device is temporarily disconnected from its circuit. After the resistance measurement is made, the circuit is restored by melting a small drop of solder across the slit.

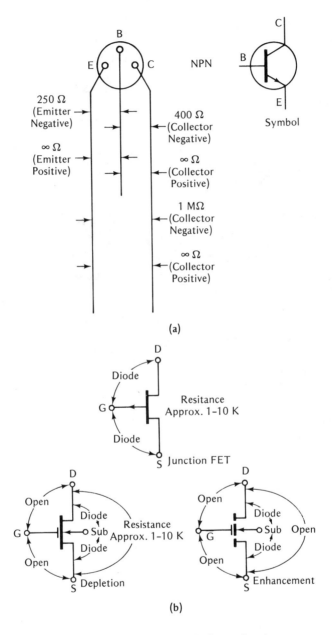

Figure 3-17. Typical junction-resistance and channel-resistance indications by a 20,000 ohms-per-volt multimeter: (a) bipolar small-signal transistor; (b) unipolar small-signal transistors.

62 RESISTANCE MEASUREMENTS

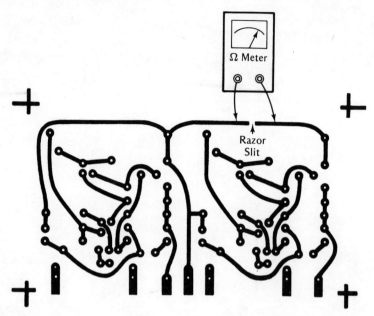

Figure 3-18. Printed-circuit conductor may be slit to make an in-circuit resistance measurement.

Review Questions

1. Distinguish between linear and nonlinear resistance.

2. What is the distinction between positive resistance and negative resistance?

3. How does incremental resistance differ from dc resistance?

4. Explain the difference between tangible resistance and intangible resistance.

5. Describe the basic distinction between a conventional ohmmeter and a low-power ohmmeter.

6. Discuss how a nonlinear resistance measurement should be specified.

7. What is the basic characteristic of a thermistor?

8. How does a light-dependent resistor differ from a wire-wound resistor?

9. Explain how generator output resistance is measured.

10. Describe the measurement of amplifier input resistance.

11. Discuss the measurement of amplifier output resistance.

12. Define a resistance strain gage. *P. 57*

13. How can the internal resistance of a battery be measured?

14. State a typical junction-resistance value.

15. Describe an expedient that may be employed for in-circuit resistance measurement.

chapter four

DC Voltage Measurements

4-1 General Considerations

DC voltage measurements are basically straightforward, although good practices should be observed. The necessity for avoiding objectionable circuit loading has been noted. Reverse polarities can be encountered in transistor circuitry (see Figure 4-1). If the pointer deflects off-scale to the left, the operator may merely reverse the test leads, provided that a multimeter is being used. If the operator happens to be using an old-style vacuum-tube voltmeter (VTVM), however, it is not permissible to reverse the test leads. Because the ground lead of a VTVM has very low input impedance, it will load the circuit excessively if it is used as a "hot" test lead. Some old-style VTVM's also are subject to instrument-circuit disturbance under this condition, and will indicate 150 V for example, when the test leads are connected to a 3-volt potential. To avoid this trouble, the operator should throw the polarity switch of the VTVM to obtain an up-scale reading, instead of reversing the test leads.

Note also in Figure 4-1 that it is good practice to set the range switch to a much higher position than would seem to be needed. After the test leads are applied, the range switch can then be set to a lower position as may be required. This precaution is particularly important when an old-style multimeter is being used—these meters were not provided with overload protec-

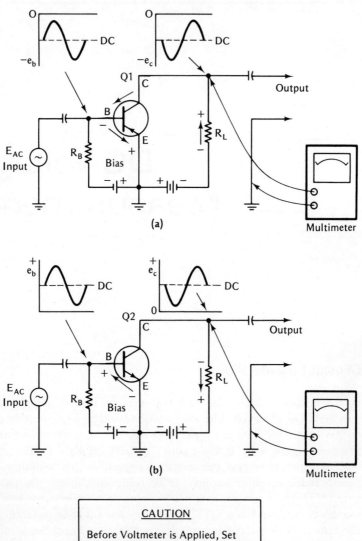

Figure 4-1. Example of reverse polarities in similar circuits: (a) PNP transistor operates with negative collector voltage; (b) NPN transistor operates with positive collector voltage.

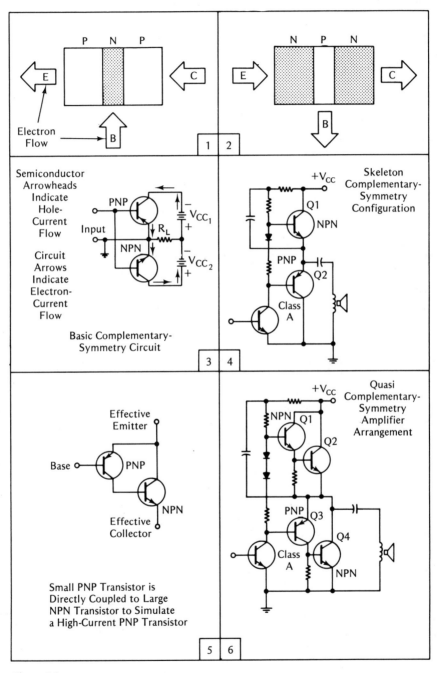

Chart 4-1.

68 DC VOLTAGE MEASUREMENTS

tion and a wrong "guesstimation" will result in serious damage to the meter movement. When electrical or electronic equipment becomes defective, unexpectedly high voltages may be encountered at points that normally operate at low potentials.

Electron Current and Hole-Current Flow

When polarities of dc voltages are considered in semiconductor circuitry, hole-current flow may be discussed in addition to electron-current flow. From a practical viewpoint, a hole is regarded as a positive electron. *Hole-current flow occurs only within semiconductor substances.* All currents in external circuitry are electron currents. Hole-current flow is essentially the same as conventional-current flow—its direction is opposite to that of electron-current flow. With reference to Chart 4-1, note that the arrowhead in a diode or transistor symbol points in the direction of hole flow (opposite to the direction of electron flow). The arrowhead points in the direction of conventional-current flow.

4-2 DC Voltage Distribution in Transistor Circuitry

A PNP transistor normally operates with its collector negative in respect to its emitter. Conversely, an NPN transistor normally operates with its collector positive with respect to its emitter. In the first analysis, transistor networks can be classified into two-battery arrangements and one-battery arrangements, as shown in Figure 4-2. These are the basic class-A amplifier arrangements. Each of the transistors is forward-biased. Observe that a PNP transistor is forward-biased when its base is negative with respect to its emitter. On the other hand, a NPN transistor is forward-biased when its base is positive with respect to its emitter. In these particular examples, the emitter of each transistor is grounded (the collector may be grounded in a variation of the circuit). In turn, the collectors of the illustrated PNP transistors are negative *with respect to ground*; the collectors of the NPN transistors are positive with respect to ground. Observe that the bases of the PNP transistors in this example are negative with respect to ground while the bases of the NPN transistors are positive with respect to ground.

Since the batteries are returned to the emitter terminals of the transistors depicted in Figures 4-2(a) and (b), these configurations are called common-emitter (CE) circuits. Similarly, because the battery is returned to the emitter terminal in Figures 4-2(c) and (d), the configurations also are called common-emitter circuits. Moreover, inasmuch as the emitter terminals are grounded, these arrangements are also examples of grounded-emitter circuits. Note carefully that the term *common ground* should not be confused with the term *common emitter*. In other words, a common ground is

4-2 DC VOLTAGE DISTRIBUTION IN TRANSISTOR CIRCUITRY 69

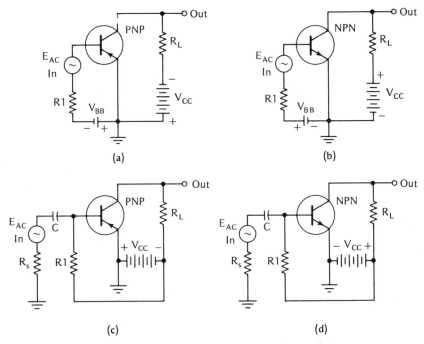

Figure 4-2. Terminal polarities in the common-emitter configuration: (a) two-battery PNP transistor circuit; (b) two-battery NPN transistor circuit; (c) one-battery PNP transistor circuit; (d) one-battery NPN transistor circuit.

the point in a semiconductor circuit that is connected to the chassis of the equipment or to an equivalent ground bus. On the other hand, a common-emitter circuit is one in which the voltage source(s) for collector and base are returned to the emitter terminal.

Measurement of dc Voltage with ac Component

Observe in Figure 4-1 that if input ac voltage E_{AC} is applied, that ac voltage is superimposed on the dc voltages at the base and emitter terminals. Since dc voltmeters are designed to measure dc voltage in the presence of ac voltage, the same value of dc voltage will be indicated whether the ac voltage is present or absent. This independence assumes, however, that the ac voltage is not excessive. If the transistor is overdriven by the ac voltage, rectification will occur in addition to amplification, and the dc levels at the base and emitter terminals will be shifted. In other words, it is permissible to check dc voltage levels in an amplifier provided that the transistor is operating in class A. Note in passing that receiver service data usually specifies dc voltage values with no signal present.

70 DC VOLTAGE MEASUREMENTS

Next, refer to Figure 4-3. The batteries are returned to the base terminal of the transistor in the common-base (CB) configuration. In the examples of Figures 4-3(c) and (d), the V_{cc} return is not made directly to the base terminal but via resistors $R2$ and $R3$. In each of the examples in Figure 4-3, we will observe that the input ac voltage is applied effectively between emitter and base, and that the output ac voltage is taken effectively between collector and base. These particular *common-base circuits* are also examples of *grounded-base circuits*. Note that the PNP transistors operate with base positive and collector negative, whereas the NPN transistors operate with base negative and collector positive.

Another basic mode of bipolar transistor operation employs the *common-collector* (CC) configuration, depicted in Figure 4-4. Supply-voltage return is made to the collector in the CC arrangement. Since the collector terminal is grounded in these particular examples, they also typify *grounded-collector* configurations. The ac input voltage is applied effectively between base and collector, and the output ac voltage is taken effectively between emitter and collector. The PNP transistors operate with base negative with respect to emitter, and with collector negative with respect to emitter. On

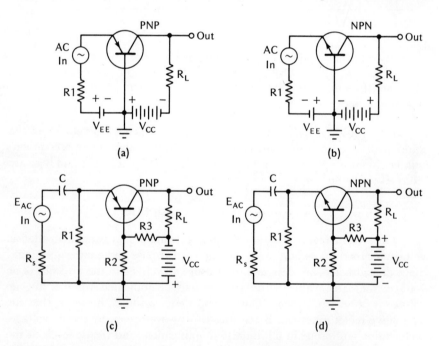

Figure 4-3. Terminal polarities in the common-base configuration: (a) two-battery PNP transistor circuit; (b) two-battery NPN transistor circuit; (c) one-battery PNP transistor circuit; (d) one-battery NPN transistor circuit.

4-2 DC VOLTAGE DISTRIBUTION IN TRANSISTOR CIRCUITRY

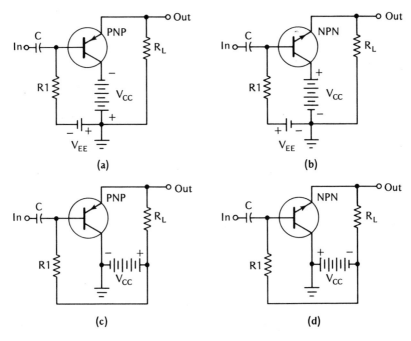

Figure 4-4. Supply-voltage polarities in the common-collector configuration: (a) two-battery PNP transistor circuit; (b) two-battery NPN transistor circuit; (c) one-battery PNP transistor circuit; (d) one-battery NPN transistor circuit.

the other hand, the NPN transistors operate with base positive with respect to emitter and with collector positive with respect to emitter. In summary, *polarities of dc voltages between base, emitter, and collector terminals are the same, whether the CE, CB, or CC configuration is employed.*

Unipolar Transistor Configurations

Next, observe the symbols for various types of field-effect (unipolar) transistors shown in Figure 4-5. Depletion types are in most general use; the enhancement type of FET is utilized chiefly in digital equipment. There are three basic FET configurations, as shown in Figure 4-6; these are the common-source, the common-gate, and the common-drain arrangements. Since the source terminal is grounded in (a), this is also called a grounded-source circuit. Similarly, since the gate terminal is grounded in (b), this configuration is also termed a grounded-gate circuit. The drain is effectively grounded in (c), and this configuration is also called a grounded-drain arrangement. N-type depletion FET's are illustrated and operate with the battery polarities indicated.

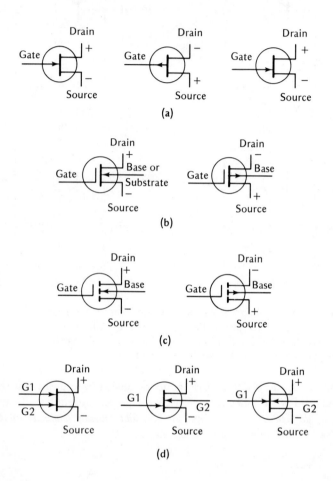

Figure 4-5. Symbols for various types of FET's: (a) JFET's—arrow points to N-type substance, away from P-type substance; (b) depletion-type MOSFET's—arrow points to N-type substrate, away from P-type substrate; (c) enhancement-type MOSFET's—arrow points to N-type substrate, away from P-type substrate; (d) dual-gate FET's—N-channel symmetrical type is at left—N-channel nonsymmetrical type is at center—alternate symbol is at right.

4-2 DC VOLTAGE DISTRIBUTION IN TRANSISTOR CIRCUITRY

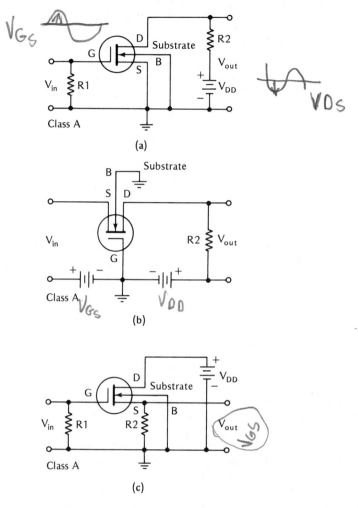

Figure 4-6. Three basic FET configuration: (a) common-source; (b) common-gate; (c) common-drain.

Voltages with Respect to Ground

Consider next the basic examples of voltage polarities with respect to ground, as depicted in Figure 4-7. As noted previously, a CE configuration does not necessarily employ a grounded-emitter connection. In the example of Figure 4-7(a), the emitter terminal is indirectly grounded via V_{cc}. Accordingly, all of the circuit's dc voltages are positive with respect to ground. Observe that the collector potential is positive with respect to ground, but it is negative with respect to the emitter terminal. Similarly, the base potential

74 DC VOLTAGE MEASUREMENTS

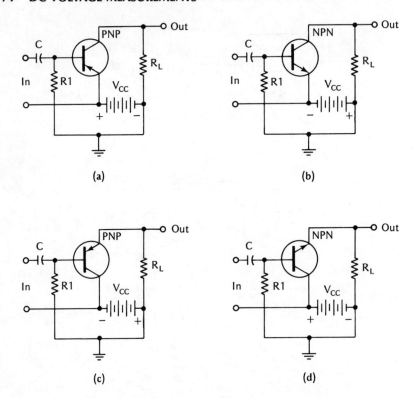

Figure 4-7. Examples of voltages with respect to ground: (a) PNP transistor in CE configuration, negative terminal of battery grounded—all voltages are positive with respect to ground; (b) NPN transistor in CE configuration, positive terminal of battery grounded—all voltages are negative with respect to ground; (c) PNP transistor in CC configuration, positive terminal of battery grounded—all voltages are negative with respect to ground; (d) NPN transistor in CC configuration, negative terminal of battery grounded—all voltages are positive with respect to ground.

is negative with respect to the emitter terminal. In (b), an NPN transistor is shown in the CE mode with the positive terminal of V_{cc} grounded (the emitter is indirectly grounded via V_{cc}). Consequently, all of the circuit voltages are negative with respect to ground. However, the collector and base potentials are positive with respect to the emitter terminal.

Refer to Figure 4-7(c). Here, a PNP transistor operates in the CC mode with the positive terminal of V_{cc} grounded (the collector is indirectly grounded via V_{cc}, or, the collector is at ac ground potential). Accordingly, all of the circuit potentials are negative with respect to ground. Also, the base and collector voltages are negative with respect to the emitter terminal.

4-2 DC VOLTAGE DISTRIBUTION IN TRANSISTOR CIRCUITRY

Next, in (d), an NPN transistor operates in the CC mode with the negative terminal of V_{cc} grounded. As in the foregoing example, the collector is indirectly grounded via V_{cc}, or, the collector is grounded for ac. All of the circuit potentials are positive with respect to ground. Also, the base and collector voltages are positive with respect to the emitter terminal.

Bias Voltage Relations

Observe the dc voltage distributions shown in Figure 4-8. A germanium transistor operating in a class-A amplifier circuit employs a forward-bias potential of typically 0.2 V. With reference to (a), note that the negative terminal of V_{cc} is grounded. In turn, all voltages in this circuit are positive with respect to ground. Because the emitter is more positive than the base, the transistor is forward-biased. This value of forward bias is equal to the difference between 2.4 and 2.2 V, or 0.2 V. Next, with reference to (b), the transistor's terminal voltages are negative with respect to the emitter (are

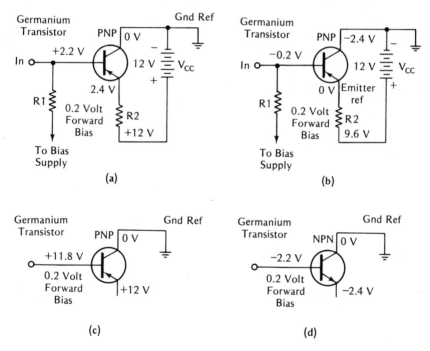

Figure 4-8. Examples of dc voltage distributions: (a) circuit voltages measured with respect to ground; (b) same circuit voltages as in (a), but measured with respect to emitter; (c) emitter is more positive than base—transistor is forward-biased; (d) emitter is more negative than base—transistor is forward-biased.

76 DC VOLTAGE MEASUREMENTS

less positive than the emitter voltage). This principle of voltage distribution is again exemplified in (c); in this case, the emitter and base voltages are much higher than the collector voltage, but their difference is 0.2 V, and the transistor is forward-biased. Finally, a forward-biased NPN transistor is depicted in (d), with its emitter terminal at a higher negative potential than its base terminal.

A bias voltage can be determined by subtracting the emitter voltage value from the base-voltage value in Figure 4-8(c). However, this method is less accurate and is more time-consuming than if the base-emitter voltage is measured directly, as depicted in Figure 4-9. Although it is often informative to determine the emitter-ground voltage, the first and most important step in troubleshooting is to check the base-emitter bias voltage. The bias voltage is far more critical than the emitter-ground voltage. If the bias voltage is checked as shown in Figure 4-9, the voltmeter can be operated on a lower range and a more accurate reading is usually obtained. Moreover, when the bias voltage is determined by taking the difference between the base and emitter voltage measurements, the percentage of error in the answer is likely to be very much greater than the error in either one of the measurements.

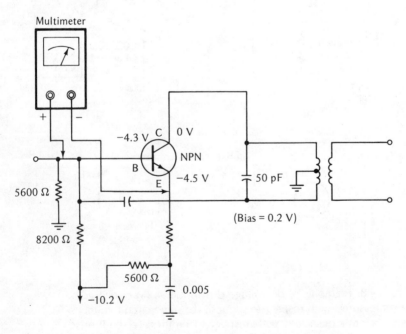

Figure 4-9. Measurement of difference voltage in a bias circuit.

4-3 DC Voltage Changes From Transistor Defects

Defective transistors are often pinpointed as a result of changes in dc voltage distribution. For example, Figure 4-10(a) shows a normal dc voltage distribution for a PNP silicon transistor circuit. The change in voltage distribution caused by an open collector junction is shown in (b). In this situation, the

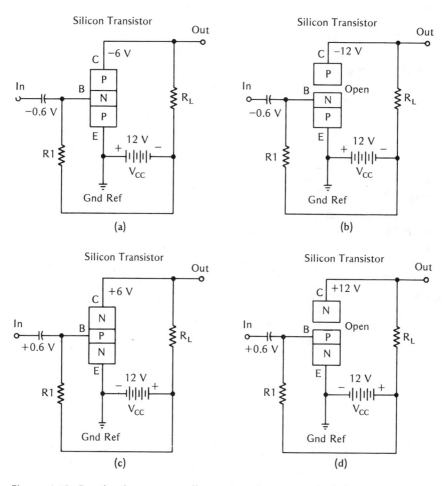

Figure 4-10. Result of an open collector junction: (a) typical dc voltages for a silicon transistor; (b) collector voltage rises to V_{cc} when collector junction is open; (c) typical dc voltages for a silicon NPN transistor; (d) collector voltage rises to V_{cc} when collector junction is open.

78   DC VOLTAGE MEASUREMENTS

collector voltage rises to the V_{cc} value, but the base-emitter bias voltage remains virtually unchanged. Figure 4-10(c) shows a normal distribution of dc voltages in an NPN silicon transistor circuit; Figure 4-10(d) shows the change in voltage distribution resulting from an open collector junction. As in the previous example, the collector voltage rises to the V_{cc} value and the base-emitter bias voltage remains virtually unchanged. Note that a silicon transistor has a typical forward bias of 0.6 V for class-A operation. This is approximately twice the bias voltage for a germanium transistor in class-A operation.

Although the collector voltage of a transistor is not critical, the base-emitter bias voltage is subject to a tight tolerance. For example, consider a transistor amplifier stage that normally operates at a collector potential of 12 V. If the supply voltage decreases to 9 V (a 25 percent change), stage operation will not be greatly affected, although the transistor will start to overload and distort at a lower signal level than otherwise. On the other hand, suppose that the bias voltage on the transistor decreases by 25 percent, as seen in Figure 4-11. If the base-emitter bias voltage decreases from 0.2 to 0.15 V, the transistor will be almost cut off, and the output signal will be very weak. Therefore, the bias voltage is checked first in case of stage malfunction.

Next, consider the changes in dc voltage distribution caused by an open emitter junction. A normal dc voltage distribution for a germanium PNP transistor is depicted in Figure 4-12(a), and the corresponding change in voltage distribution owing to an open emitter junction is shown in (b). In this fault situation, both the collector and base voltages rise to the V_{cc} potential. A normal dc voltage distribution for a germanium NPN transistor is depicted in (c), and the corresponding change in voltage distribution

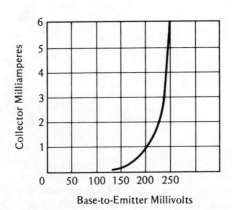

Figure 4-11. Transfer characteristic for a small-signal germanium transistor.

4-3 DC VOLTAGE CHANGES FROM TRANSISTOR DEFECTS 79

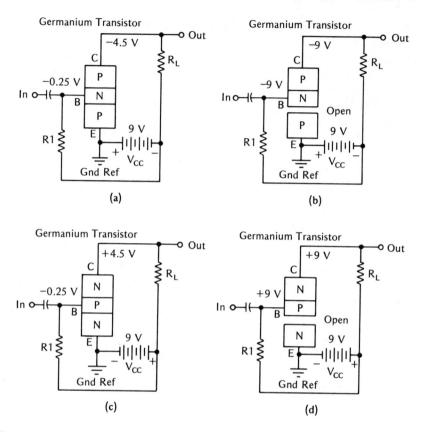

Figure 4-12. Result of an open emitter junction: (a) typical dc voltages for a germanium PNP transistor; (b) collector and base voltages rise to V_{cc} when emitter junction is open: (c) typical dc voltages for a germanium NPN transistor; (d) same result as (b).

owing to an open emitter junction is seen in (d). As in the preceding example, the collector and base voltages rise to the V_{cc} potential. When a junction is open, there is no current flow across the junction. This fact would be immediately evident if current measurements were made in the examples of Figures 4-12(b) and (d). Professional technicians, however, prefer to employ voltage measurements in the vast majority of trouble situations. Current measurements in analog circuitry are comparatively difficult to make.

Collector Junction Leakage

Another common transistor fault is termed *leakage*, or a "leaky collector junction." In other words, the defective junction is neither open-circuited

80 DC VOLTAGE MEASUREMENTS

nor short-circuited, but resistive. This leakage resistance has the action of permitting reverse-current flow from collector to base. Stage gain decreases, and the transistor sometimes generates excessive noise. A normal dc voltage distribution for a silicon NPN transistor is shown in Figure 4-13(a). The corresponding change in voltage distribution owing to a leaky collector junction is shown in (b). Junction leakage causes the base-emitter forward bias to increase and causes the collector-emitter voltage to decrease. Since the forward-bias voltage becomes excessive, the current flow becomes excessive. In turn, most of the V_{cc} voltage appears as an abnormal IR drop across R_L. The corresponding normal and abnormal dc voltage distributions for a silicon PNP transistor are depicted in Figures 4-13(c) and (d).

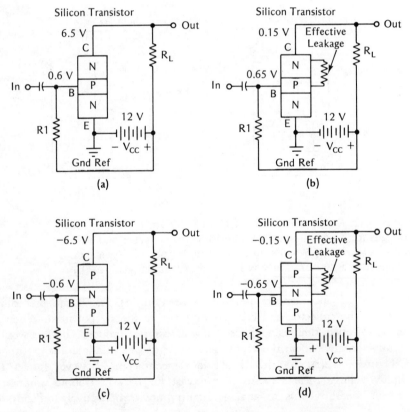

Figure 4-13. Result of a leaky collector junction: (a) typical dc voltages for a silicon NPN transistor; (b) collector voltage decreases and base voltage increases when collector junction is leaky; (c) typical dc voltages for a silicon PNP transistor; (d) collector voltage decreases and base voltage increases when collector junction is leaky.

4-3 DC VOLTAGE CHANGES FROM TRANSISTOR DEFECTS   81

Sometimes an emitter junction becomes short-circuited. When this happens, the base voltage assumes the same value as the emitter voltage (zero in the example of Figure 4-14). Effectively, the transistor has been changed into a reverse-biased diode by the fault and the collector voltage rises to the V_{cc} potential. There is current flow through $R1$ via the short-circuited emitter junction to ground. However, this current cannot branch into the collector circuit because the emitter junction is lacking. A somewhat similar abnormal dc-voltage distribution can be caused by an open connection to the base terminal. The measured base voltage will either be equal to V_{cc} or will be a small positive voltage, depending upon which end of the open connection is being tested. Since the base will be "floating" in this situation, the base-voltage value that is indicated will depend considerably upon the internal resistance of the voltmeter. A low ohms-per-volt meter will indicate practically zero volts.

Aside from the possibility of false indication owing to circuit loading, a low ohms-per-volt meter occasionally causes puzzling circuit responses when voltages are checked in open-circuit situations. In Figure 4-15, the open-source condition results in a "dead" stage. However, when a low ohms-per-volt meter is applied to check the source voltage, it is likely that the stage will resume operation and the source voltage may measure a reasonable value. But when the test leads are disconnected, the stage is "dead" again. Of course, the unexpected circuit response is the result of temporary substitution of R_{in} for R_s. If a high ohms-per-volt meter is utilized in this situation, the stage will not resume operation and the meter will indicate the drain or supply-voltage value.

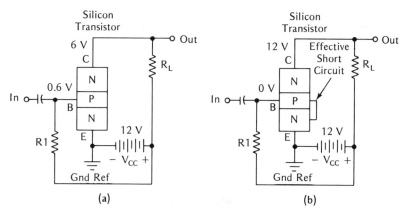

Figure 4-14. Result of a short-circuited emitter junction: (a) typical dc operating voltages for a silicon NPN transistor; (b) base voltage becomes zero and collector voltage equals V_{cc} when the emitter junction is short-circuited.

82 DC VOLTAGE MEASUREMENTS

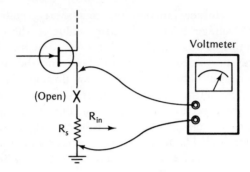

Figure 4-15. A low-sensitivity voltmeter effectively completes the source circuit.

4-4 Tolerances on DC Voltage Values

Tolerances on measured values stem from two sources. First, the voltmeter may be inaccurate. If, however, a good-quality instrument is utilized, this possibility can be dismissed. Second, all components and devices in the equipment under test have certain manufacturing tolerances. Thus, resistors may have a tolerance of ±10 percent and often ±20 percent. Resistors in critical circuits have tolerances as tight as ±1 percent. The supply voltage to the equipment is subject to drift and supply-voltage variation affects the dc-voltage distribution throughout the equipment. Semiconductor devices often have comparatively wide tolerances. It is not unusual for a replacement transistor to draw 25 percent more or less current than the design-center value. In turn, associated circuit voltages are affected.

Although circuit malfunctions are associated with abnormal dc-voltage distributions, it is seldom feasible to establish hard-and-fast limits to separate an acceptable circuit from an unacceptable circuit. Many malfunctions involve more or less impaired performance (although the performance may be deemed acceptable). Other malfunctions are associated with catastrophic failure of circuit action. Various television receiver manufacturers state that specified dc-voltage values are subject to a ±20 percent tolerance with the line voltage adjusted to standard value. However, this tolerance on dc-voltage values is merely a very general guideline.

Substantial departures from specified dc-voltage values throw suspicion on associated circuitry—they do not prove that this circuitry is in need of repair. Accordingly, additional tests and evaluations must be made before a firm conclusion can be made concerning the significance of the off-tolerance voltage values. Analysis is facilitated if few or no interacting circuit sections are involved. When sectional interactions are present in a receiver (or other) system, a malfunction in one section may be "reflected" into another section, thereby producing misleading trouble symptoms. This possibility is aggravated in circuitry that normally operates with signal-developed bias voltages. Another serious difficulty is encountered in direct-coupled circuitry. However, the problem can be reduced by employment of suitable analytical approaches as detailed subsequently.

4-5 Measurement of High DC Voltage

Most dc voltmeters are limited to a top indication in the 1-kV to 5-kV range. In turn, dc voltages up to 40 kV must be measured with the aid of an external (accessory) multiplier resistor. A high-voltage multiplier resistor is enclosed in a well-insulated housing, as shown in Figure 4-16. This arrangement is called a high-voltage probe. The required value for a multiplier resistor depends upon the input resistance of the voltmeter with which it is used. Therefore, most high-voltage probes are designed so that the multiplier cartridge can be changed as required for use with a particular meter. There has been a trend toward the use of high-voltage dc probes with self-contained meters. The probe illustrated in Figure 4-17 has a range up to 40 kV. Other top ranges are available to suit various applications.

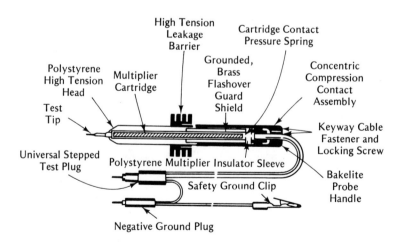

Figure 4-16. Construction of a typical high-voltage dc probe.

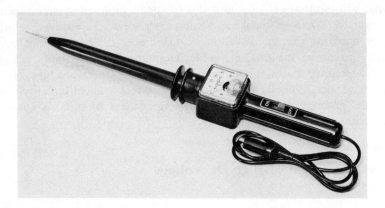

Figure 4-17. A high-voltage dc probe with self-contained meter movement. (*Courtesy*, B & K Precision, Div. of Dynascan Corp.)

Review Questions

1. Describe the technical problem of circuit loading in dc-voltage measurements.

2. Characterize hole-current flow versus electron current flow.

3. Explain the term *dc voltage distribution*.

4. Distinguish between common-emitter, common-base, and common-collector transistor configurations.

5. How should the base-emitter bias voltage of a transistor be measured?

6. Discuss how collector-junction leakage in a transistor affects the associated dc-voltage distribution.

7. Describe the function of a high-voltage dc probe.

8. Why should the ground lead of a VTVM never be utilized as a "hot" test lead?

9. Explain why it is good practice to set the range switch of a multimeter to a much higher position than would seem to be needed when an initial measurement is made.

10. Define conventional current flow.

11. Why is a common-emitter configuration not necessarily identical with a grounded-emitter configuration?

12. If a 1-volt rms potential with a sine waveform is applied to a dc voltmeter, what value does the pointer indicate?

13. How is a normal dc-voltage distribution changed by an open emitter junction in a transistor?

14. How is a normal dc-voltage distribution changed by a short-circuited emitter junction in a transistor?

15. Discuss the general nature of tolerances on dc voltage values.

chapter five

AC Voltage Measurements

5-1 General Considerations

AC voltage values are based on defined values in a sine waveform as depicted in Figure 5-1. Thus, a sine wave has a positive-peak value and an equal negative-peak value. It has a peak-to-peak value equal to the sum of its peak values. It has a rms value equal to 0.707 of its peak value. Any arbitrary point along the sine wave may be specified in terms of its instantaneous value. Most ac voltmeters indicate in rms units with respect to a sine wave. Some ac voltmeters indicate in both peak-to-peak units and rms units with respect to a sine wave. A voltmeter that indicates in peak-to-peak units occasionally employs peak-to-peak responding instrument circuitry. In such a case, the instrument will indicate the true peak-to-peak voltage of a complex waveform as well as that of a sine waveform. As noted previously, most ac voltmeters utilize instrument rectifiers that develop the average value of the rectified waveform; this type of voltmeter can indicate a rms peak, or peak-to-peak value correctly only if the input waveform is sinusoidal.

A few ac voltmeters are designed as *true-rms* instruments. This term denotes that the voltmeter is free from waveform errors and that it will indicate the true rms value of any waveform. As an illustration, an electrodynamic voltmeter that indicates true rms values of sine waves, complex waves, and dc potentials is shown in Figure 5-2. An electrodynamometer

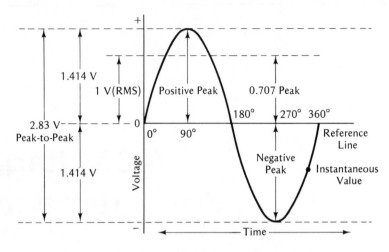

Figure 5-1. Basic values and their relations in a sine waveform.

Figure 5-2. An electrodynamic voltmeter combined with an ammeter and a wattmeter. (*Courtesy*, Simpson Electric Co.)

movement utilizes a fixed coil and a rotatable coil pivoted inside of the fixed coil. The chief disadvantage of an electrodynamic instrument is that it imposes a comparatively heavy current burden and can be applied only in low-impedance circuits. High-impedance true-rms voltmeters that employ elaborate electronic circuitry are available, although they are comparatively costly.

5-2 Measurement of AC Voltage with DC Component Present

If an ac voltmeter is used to measure the signal output from an audio-amplifier transistor, the sine-wave signal is mixed with a dc component. In other words, the operator must measure the ac component in a pulsating-dc waveform, as depicted in Figure 5-3. This measurement requires the use of a blocking capacitor in series with the instrument circuitry of the ac voltmeter, so that the dc component is prevented from flowing into the meter and thereby falsifying the reading. Many ac voltmeters contain built-in blocking capacitors, so that no precautions need be observed when making ac-voltage measurements on pulsating-dc waveforms. Some ac voltmeters do not include a blocking capacitor, and this component must then be externally provided by the operator. Most multimeters provide an ac-voltage function

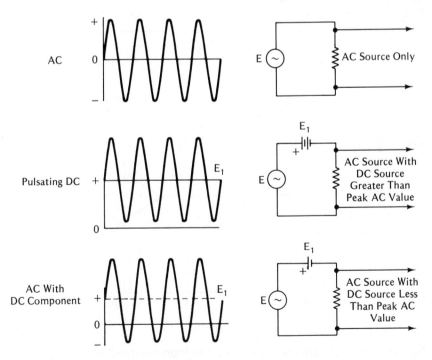

Figure 5-3. An ac waveform with three dc component levels.

90 AC VOLTAGE MEASUREMENTS

and an output function. If a dc component is present in a waveform, the operator utilizes the output function. The chief disadvantage of an output function is its inability to process low frequencies, below 60 Hz, for example.

5-3 AC-to-DC Converters

AC voltmeters in volt-ohm-milliammeters (VOM's) utilize semiconductor diodes that are called instrument rectifiers. The ac voltmeter section in a digital multimeter (DMM) often employs similar semiconductor diodes that

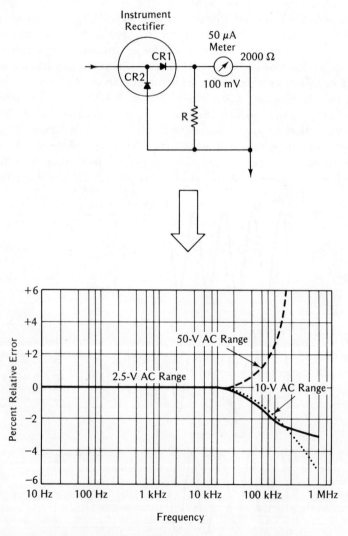

Figure 5-4. Typical frequency response of an instrument-rectifier network on its ac-voltage function.

5-4 TURNOVER ERROR IN AC VOLTAGE MEASUREMENT

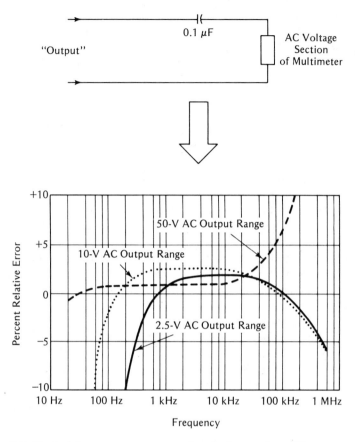

Figure 5-5. Typical frequency response of an instrument-rectifier network on its output function.

are termed ac-to-dc converters or ac converters. Frequency response is generally uniform to at least 10 kHz, as shown in Figure 5-4. Some VOM's and DMM's provide extended high-frequency response. An occasional DMM provides power-frequency response only. This type of DMM has a frequency response from 20 Hz to 1 kHz. If an output function is provided, a blocking capacitor is switched in series with the instrument rectifiers (ac converter). A typical value of blocking capacitance is 0.1 µF. As illustrated in Figure 5-5, the blocking capacitor reduces the low-frequency response of the ac voltmeter and also impairs its midrange accuracy to some extent.

5-4 Turnover Error in AC Voltage Measurement

Turnover error is a type of waveform error. Waveform error occurs when non-sinusoidal voltages are measured with instrument rectifiers, but turnover

92 AC VOLTAGE MEASUREMENTS

error is encountered only with half-wave instrument rectifiers. If the scale indication changes when the test leads to the ac voltmeter are reversed, turnover error is present. A configuration for an RF (radio frequency) probe is shown in Figure 5-6(a). This type of probe is used to make ac voltage measurements at frequencies up to 200 MHz. It is a rectifier-type probe that is connected to a dc voltmeter. The probe employs half-wave rectification and develops the peak value of the applied ac voltage. A voltage waveform with even-harmonic distortion is depicted in Figure 5-6(b). It is evident that if the probe leads are applied in one polarity, the meter response will be to V_1. If the probe is applied in the other polarity, the meter response will be to V_2. This is an example of turnover error.

The RF probe exemplified in Figure 5-6(a) contains a 4.1-MΩ series resistor. If the TVM has an input resistance of 10 MΩ, the scale indication will be in rms values of sine waves. The chief disadvantage of this type of probe is its input voltage limitation. In other words, the diode will be damaged if the probe is applied in a circuit that operates with a peak voltage in excess of the diode rating. This limitation is in the order of 20 volts for typical diodes. A minority of TVM's provide instrument circuitry or

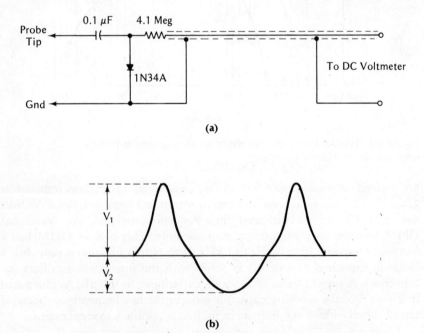

Figure 5-6. Turnover error can occur with half-wave instrument rectifier: (a) RF probe for standard TVM; (b) waveform with unequal positive and negative excursions.

accessory probes that respond to the peak-to-peak voltage of complex waveforms. This kind of instrument indicates true peak-to-peak voltage values of complex waveforms. However, it can indicate true rms values of sine waves only.

5-5 Measurement of Small AC Voltages

Conventional ac voltmeters cannot measure ac voltages in the millivolt range. However, various applications, such as audio test procedures for example, require measurements in the millivolt range. For this purpose, an audio voltmeter (sensitive ac voltmeter) such as the one illustrated in Figure 5-7 is employed. This instrument has a first range of 1 mV full scale and a top range of 300 V full scale. Frequency response is from 10 Hz to 1 MHz. Indication is in rms values of sine waves. A sensitive ac voltmeter is used to measure the output voltage from a phono cartridge, tape head, voltages in the input circuits of preamplifiers, hum voltage in V_{cc} lines, output voltage from a microphone, and similar low-level measurement requirements. This type of ac voltmeter can also be used to measure sine-wave amplitudes in

Figure 5-7. A sensitive ac voltmeter used in low-level audio test procedures. (*Courtesy,* Heath Co.)

94 AC VOLTAGE MEASUREMENTS

high-level circuitry, up to a limit of 300 V. Note that a sensitive ac voltmeter provides no functions other than that of ac voltage measurement.

5-6 Tuned AC Voltmeters

Many ac voltage measurements are made with tuned ac voltmeters. For example, a relative field-strength meter such as that illustrated in Figure 5-8 is used to measure television signal voltages on any selected channel from 54 to 890 MHz. It is essentially an all-channel TV tuner that energizes a sensitive TVM with calibration facilities. Another example of a tuned ac voltmeter is a harmonic distortion analyzer used in audio test procedures. This instrument contains a tunable filter that is adjusted to cancel out the fundamental component of the incoming audio test signal. In turn, the voltmeter indicates the effective amplitude of the harmonic components (if any) in the incoming signal. Still another example of a tuned ac voltmeter is an intermodulation analyzer, also used in audio test procedures. This instrument employs a two-tone test signal and contains both high-pass and low-pass filters with a demodulator section. The voltmeter indicates the effective amplitude of the intermodulation distortion products (if any) that are produced by the equipment under test.

Figure 5-8. Appearance of a field-strength meter. (*Courtesy,* Sencore)

5-7 True rms AC Voltage Measurement

As noted previously, specialized voltmeters are required to measure the rms value of a complex waveform. Although electrodynamic voltmeters indicate true rms values, regardless of waveform, they have the disadvantages of high current burden and of limited frequency response. To obtain high input impedance and extended high-frequency response, one type of true-rms ac voltmeter is designed with electronic amplifiers that drive a thermocouple element. In turn, the potential that is generated by the thermocouple is indicated by a dc meter movement; another version of this design employs digital readout. A thermocouple consists of two dissimilar metals joined at one end. If the junction of these metals is heated, a dc voltage is produced at the free terminals. This voltage is proportional to the heat difference between the hot and cold ends. It is characterized as a thermoelectric action.

A thermocouple meter consists of a heater through which the ac current flows, a thermocouple attached to the heater element, and a dc voltmeter connected to the free ends of the thermocouple. In an rms voltmeter, the heater is not energized directly by the ac voltage under test. Instead, the voltage to be measured is passed through an electronic power amplifier arrangement prior to energizing the heater. The dc voltage that develops at the free end of the thermocouple causes a meter indication that is directly proportional to the rms value of the ac voltage under test. In turn, the scale is calibrated in rms units. The heater is fabricated from platinum or nickel alloy and is usually in direct contact with the junction of the thermocouple. The thermocouple itself is made of constanstan and platinum alloy.

The thermocouple develops a dc voltage in response to the temperature difference between the hot (junction) and the cold (free) ends. This temperature difference should be caused only by the current being measured and should not be influenced by atmospheric (ambient) temperature. To eliminate this source of error, the free ends of the thermocouple should be attached to the center of separate copper strips, so that they will be constantly at the average temperature of the ends (not the center) of the heater element. The ends of the copper strips are placed close enough to the heater ends to have the same temperature, but are electrically insulated from them by means of thin mica sheets, as seen in Figure 5-9. When current flows, the temperature at the center of the heater element becomes much greater than the temperature at the heater ends or at the free ends of the thermocouple. For a given current, the temperature difference between the thermocouple junction and the free ends is constant, regardless of the ambient temperature.

96 AC VOLTAGE MEASUREMENTS

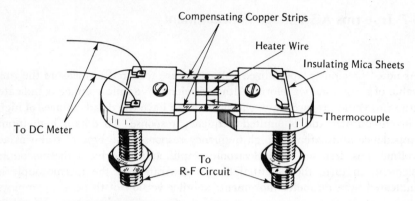

Figure 5-9. Construction of a thermocouple.

5-8 Decibel Measurements with AC Voltmeters

Most conventional ac voltmeters have decibel scales. Decibel meters such as those illustrated in Figure 5-10 are essentially rectifier-type ac voltmeters with a scale calibrated in dB units. The decibel unit is a power ratio function by definition. In turn, voltage ratios correspond to power ratios, if the two voltages are measured across equal load resistances. However, if the load values are unequal, the measured voltage ratios do not correspond directly to power ratios. Decibel values are equal to twenty times the logarithm of voltage ratios. They are additive and subtractive. (Decibel

Figure 5-10. Typical rectifier-type decibel meter. (*Courtesy, Simpson Electric Co.*)

5-8 DECIBEL MEASUREMENTS WITH AC VOLTMETERS

values are also equal to ten times the logarithm of power ratios.) As an illustration, if there is a 20-dB loss in a volume control followed by a gain of 30 dB in an amplifier, the overall gain is 10 dB.

All dB values are referenced to some stipulated power value, to which the value 0 dB is assigned. Some dB scales are referenced to 6 mW in 500 Ω (Figure 5-10), and other dB scales are referenced to 1 mW in 600 Ω. Still other reference levels will be encountered on occasion. Note that the dB scale on an ac voltmeter indicates dB values directly only when the measurements are made across a value of load resistance to which the scale is referenced. However, calculations can always be made to correct indicated dB values both for nonstandard load resistances and/or some other 0-dB reference level. A simple example of dB measurements with ac voltmeters designed to indicate dB values across 600-Ω loads is shown in Figure 5-11. A good sine waveform must be utilized in dB measurements of this sort, or the measured values will be subject to waveform error. In this example, the audio oscillator applies a signal level of 2 dB to the 600-Ω volume control, and the amplifier increases the signal level to 12 dB. Thus, the overall gain is equal to the difference between these two values, or 10 dB.

Most ac voltmeters have only one dB scale, which is usually direct-reading on the lowest ac-voltage range of the instrument. Hence, when the meter is operated on a higher ac-voltage range, a suitable number of dB must be added to the scale reading. Scale factors may be printed on the meter scale plate or they may be listed in the instruction manual for the instrument. Note that, from a practical viewpoint, although two meters are shown in Figure 5-11, only one is needed. The operator simply moves the test leads from the input terminals to the output terminals of the equipment

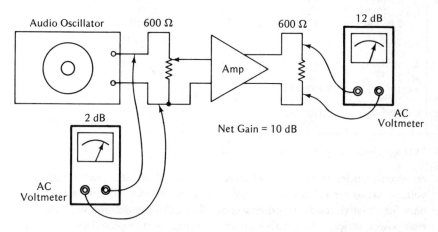

Figure 5-11. Basic example of dB measurements across 600-ohm resistive loads.

Table 5-1
Power Ratios, Voltage Ratios, and dB Values*

Power Ratio	Voltage Ratio	dB − ←	dB + →	Voltage Ratio	Power Ratio
1.000	1.0000	0		1.000	1.000
0.9772	0.9886	0.1		1.012	1.023
0.9550	0.9772	0.2		1.023	1.047
0.9333	0.9661	0.3		1.035	1.072
0.9120	0.9550	0.4		1.047	1.096
0.8913	0.9441	0.5		1.059	1.122
0.8710	0.9333	0.6		1.072	1.148
0.8511	0.9226	0.7		1.084	1.175
0.8318	0.9120	0.8		1.096	1.202
0.8128	0.9016	0.9		1.109	1.230
0.7943	0.8913	1.0		1.122	1.259
0.6310	0.7943	2.0		1.259	1.585
0.5012	0.7079	3.0		1.413	1.995
0.3981	0.6310	4.0		1.585	2.512
0.3162	0.5623	5.0		1.778	3.162
0.2512	0.5012	6.0		1.995	3.981
0.1995	0.4467	7.0		2.239	5.012
0.1585	0.3981	8.0		2.512	6.310
0.1259	0.3548	9.0		2.818	7.943
0.10000	0.3162	10.0		3.162	10.00
0.07943	0.2818	11.0		3.548	12.59
0.06310	0.2512	12.0		3.981	15.85
0.05012	0.2293	13.0		4.467	19.95
0.03981	0.1995	14.0		5.012	25.12
0.03162	0.1778	15.0		5.623	31.62
0.02512	0.1585	16.0		6.310	39.81
0.01995	0.1413	17.0		7.079	50.12
0.01585	0.1259	18.0		7.943	63.10
0.01259	0.1122	19.0		8.913	79.43
0.01000	0.1000	20.0		10.000	100.00
10^{-3}	3.612×10^{-2}	30.0		3.612×10	10^3
10^{-4}	10^{-2}	40.0		10^2	10^4
10^{-5}	3.162×10^{-3}	50.0		3.162×10^2	10^5
10^{-6}	10^{-3}	60.0		10^3	10^6
10^{-7}	3.162×10^{-4}	70.0		3.162×10^3	10^7
10^{-8}	10^{-4}	80.0		10^4	10^8
10^{-9}	3.162×10^{-5}	90.0		3.612×10^4	10^9
10^{-10}	10^{-5}	100.0		10^5	10^{10}

*Voltage ratios based on equal load resistances.

or system under test. Basic relations among dB values, power ratios, and voltage ratios are tabulated in Table 5-1. Observe that dB values are defined only for resistive loads. In other words, dB values are defined with respect to real-power ratios. Thus, although the arrangement depicted in Figure 5-12 may appear to provide a gain of 10 dB, it represents a loss of 2 dB.

When dB values are measured across various values of load resistance,

5-8 DECIBEL MEASUREMENTS WITH AC VOLTMETERS   99

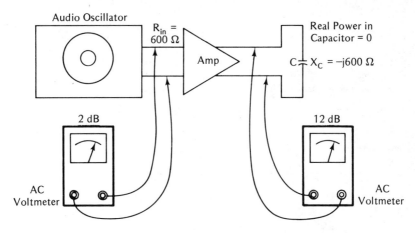

Figure 5-12. A loss of 2 dB occurs in this arrangement.

the voltage values do not correspond directly to the dB values. For example, consider the two amplifier systems depicted in Figure 5-13. In part (a), signal-voltage levels and corresponding dB values are shown for a multistage system in which all of the loads have the same value (2000 Ω). Next, in part (b), the same signal-voltage levels are shown for a similar system which differs in that successive stages have unequal load-resistance values. The result is that the measured dB values through the system are different, although all of the voltage values are the same. This example illustrates the fact that ac voltage values have no direct correspondence to dB values, and that the load values must be taken into account to determine dB values.

Two basic situations come into consideration in the measurement of dB values across resistive loads that have different values from that for which the ac voltmeter's dB scale has been calibrated. In the first instance, the load resistances are different from the reference value for the meter, but these load resistances are equal in value, as illustrated in Figure 5-14. In this situation, the operator can determine the dB gain or loss of the system by merely taking the difference between the two readings. As an illustration, the reference value for the dB scale on the ac voltmeter might be 600 Ω. If the input and load resistances of an amplifier were 75 Ω, for example, the operator can still use the meter that has a 600-Ω reference value. Note that in Figure 5-14(a), neither the 3- nor the 13-dB reading is correct in itself. On the other hand, the difference between these two readings, 10 dB, is correct inasmuch as these readings were made across equal load resistances. Therefore, the gain of the system in this example is 10 dB.

Next, observe that in Figure 5-14(b) the dB measurements are made across unequal load resistances. Neither of these resistance values corresponds to the 600-Ω reference value of the ac voltmeter with the dB scale. In

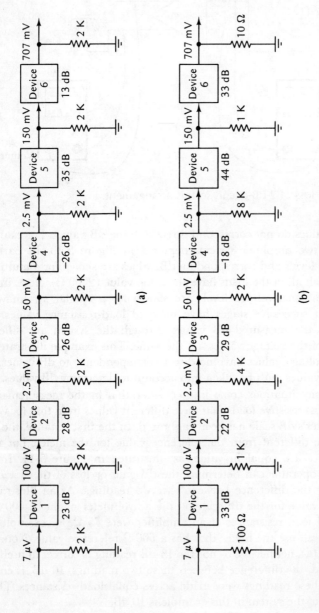

Figure 5-13. Typical changes in dB values with changes in load-resistance values; (a) voltage levels and dB levels with equal load resistances throughout; (b) voltage levels and dB levels with different values of load resistance.

5-8 DECIBEL MEASUREMENTS WITH AC VOLTMETERS 101

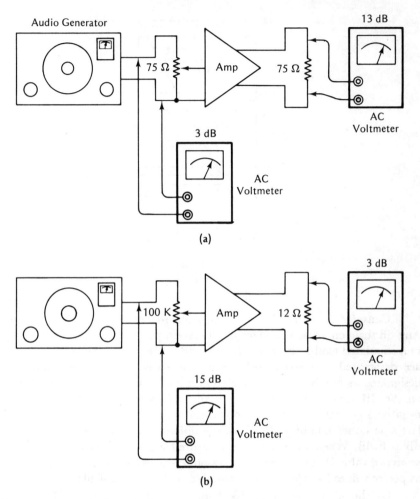

Figure 5-14. Examples of dB measurements: (a) across equal resistances; (b) across unequal resistances.

turn, neither the 15- nor the 3-dB reading is correct in itself, nor is the difference between these two readings correct. We will find that this system provides gain, although the preliminary dB readings seem to indicate that there is a loss. To evaluate these dB readings, they must be converted to suitable reference-resistive values. In other words, a corrective factor must be applied to each of the readings to determine the true dB values. The chart shown in Chart 5-2 is suitable for this purpose. If the first measurement is made across 100 kΩ, and the second measurement is made across 1000 Ω, the resistance ratio is 100. In turn, the chart shows that 20 dB would have to be added (or subtracted) to the apparent gain or loss that is indicated by the preliminary measurements.

102 AC VOLTAGE MEASUREMENTS

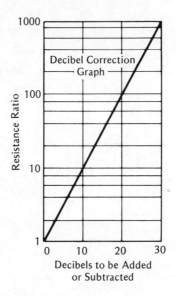

Chart 5-2.

Consider next the distinction between *relative* and *absolute* dB values. Any dB scale on an ac voltmeter has a certain power value assigned as 0 dB in a stipulated load-resistance value. All dB values below this reference level are designated as negative dB values; all above this reference level are designated as positive dB values. Red and black numerals are often utilized on the dB scale to call attention to this difference in signs. Positive and negative signs must be taken into account when the operator compares dB levels on either side of 0 dB. Thus, the difference between −5 dB and +3 dB is 8 dB. When dB values are measured across a resistor that has the reference value for the meter, these absolute values can be evaluated directly as power values by reference to a suitable table (or by calculation).

On the other hand, when dB values are measured across a resistor that has a value other than the meter's reference value, these measured values are only indirectly related to power values. It was previously noted that if the operator takes a pair of dB readings across 1000-Ω loads with a 600-Ω dB meter, the difference between these readings will be the true difference in dB units between the two ac voltage levels, although neither of the readings is correct in itself. Such readings are called *relative* dB values.

DBm Values

Now, consider the dBm unit and its measurement. This is an abbreviation for the number of decibels above (or below) a power level of 1 mW in 600 Ω. In turn, a dBm value corresponds to a quantity of power in 600 Ω expressed in terms of its ratio to 1 mW. The graph shown in Chart 5-3

5-8 DECIBEL MEASUREMENTS WITH AC VOLTMETERS 103

Resistive Load Value	DBM*
600	0
500	+0.8
300	+3.0
250	+3.8
150	+6.0
50	+10.8
15	+16.0
8	+18.8
3.2	+22.7

*DBM is the Increment to be Added Algebraically to the DBM Value Read from the Graph.

Figure 5-15. List of dBm correction factors for Figure 5-8.

indicates relations between dBm units and ac volts across 600 Ω. Zero dBm represents a power level of 1 mW in a 600-Ω resistive load; 10 dBm indicates a power level of 10 mW, and so on. If an ac voltmeter with a dBm scale is used to measure dBm values across a resistive load other than 600 Ω, the correction factors in Figure 5-15 must be applied to find the true dBm value. As noted in Chart 5-3, 0 dBm corresponds to 0.775 V across 600 Ω; similarly, 20 dBm corresponds to 7.75 V across 600 Ω.

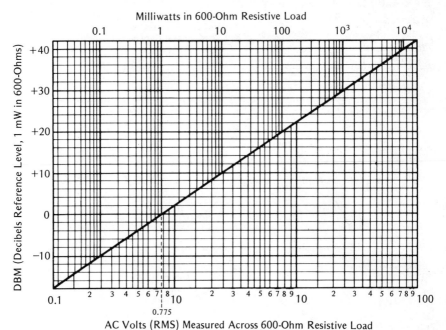

Chart 5-3.

5-9 Measurement of Negative Impedance

An air-core coil is a linear inductance. In other words, its inductance value remains constant regardless of the amount of current that flows through the coil. An iron-cored coil is a nonlinear inductance—particularly as the region of core saturation is approached. The inductance value of an iron-cored coil decreases as the core becomes progressively saturated. It follows that, if an iron-cored coil is utilized in a series LCR circuit, the resonant frequency of the circuit will shift when the current value is changed. This frequency shift is the basis of all ferroresonant circuits, and a nonlinear circuit of this kind exemplifies negative impedance. Observe the LCR series circuit depicted in Figure 5-16. Inductor L has an iron core. If the cross-sectional area of the core is comparatively small with respect to the current value I, its inductance value decreases as I increases. Otherwise stated, L operates as a saturable reactor.

Consider that the frequency of voltage E is set to a value such that for small current values, L and C resonate at a somewhat lower frequency. Then, let the current value be increased, while the operator observes the voltage measured across L and C (E_{LC}). At first, the value of E_{LC} rises as the

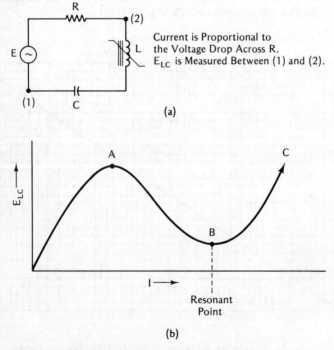

Figure 5-16. An LCR negative-impedance circuit.

5-9 MEASUREMENT OF NEGATIVE IMPEDANCE 105

current is increased. The value of L is also decreasing, however, with the result that the circuit approaches a resonant state. In turn, the voltage drops across L and C as they become more nearly equal; these voltages are 180 degrees out of phase with each other. Therefore, as the current value increases, E_{LC} will pass through a maximum value as indicated at A in Figure 5-16(b). As the current is further increased, E_{LC} passes through a minimum value at point B (the series-resonant point). Note that E_{LC} does not fall completely to zero at resonance owing to the winding resistance (R) of L. As the current is further increased, E_{LC} again rises to point C.

A ferroresonant circuit has a reactance variation that is *inductive* for small current values, and that becomes *capacitive* for high current values. This change in reactance occurs despite the fact that the driving frequency remains constant. As shown in Figure 5-16(b), the current lags the applied voltage to the left of B, but the current leads the applied voltage to the right of point B. The current and voltage are in phase (circuit is resistive) at point B. As the current is increased from a small value, E_{LC} also increases in value. On the other hand, over the interval from A to B, increasing current is accompanied by decreasing voltage; or, the current decreases as the voltage increases. In other words, the interval from A to B defines a negative impedance, but the impedance is positive from 0 to A, and from B to C.

This example of negative impedance should not be confused with negative resistance. The negative impedance is stable under any condition of operation and is solely a power sink—it cannot be operated as a power source. All of the power supplied by the source is dissipated by R, and R is always positive. In other words, the negative impedance of the characteristic from A to B in Figure 5-16(b) is developed by nonlinear reactance—not by negative resistance. A reactance cannot operate as a power sink, nor can it operate as a power source. In summary, the term negative impedance denotes only that increasing voltage values are accompanied by decreasing current values over a particular interval of the IE_{LC} characteristic.

AC VOLTAGE MEASUREMENTS

Review Questions

1. Define an instantaneous value with reference to a sine wave.

2. How is an electrodynamic voltmeter constructed?

3. Discuss three basic dc component levels in a sine waveform.

4. Explain how an "output" function is provided in an ac voltmeter.

5. Describe turnover error in ac voltage measurements.

6. What is the distinction between an audio voltmeter and a power-type ac voltmeter?

7. How does a tuned ac voltmeter operate?

8. Why does a thermocouple type of meter indicate true rms values?

9. Explain the distinction between relative and absolute dB values.

10. Discuss the role of load-resistance values in evaluation of dB scale indications by ac voltmeters.

11. Define dBm values.

12. How do departures from sinusoidal waveform affect dB indications of ac voltmeters?

13. Are dB values referenced to real power, to reactive power, or to a combination of both?

14. Discuss the relation of ac voltage ratios to ac power ratios in calculation of dB values with respect to a standard load resistance.

15. What is the standard load resistance for dBm measurements?

chapter six

Oscilloscope Tests and Measurements

6-1 General Considerations

An oscilloscope is a form of voltmeter that displays a varying voltage as a function of time. It is more informative than a basic voltmeter. An oscilloscope such as the one illustrated in Figure 6-1 provides measurement of rise time, waveform period, pulse width, duty cycle, repetition rate, peak voltage, peak-to-peak voltage, frequency, phase, percentage modulation, damping time, and various other electrical parameters. Oscilloscopes are basically classified as free-running or triggered-sweep types. An oscilloscope with free-running sweeps is adequate for visual-alignment procedures, or for audio signal tracing applications. On the other hand, pulse rise-time measurements or pulse-width measurements require the availability of triggered-sweep operation with calibrated time-base action, plus a delay section in the vertical-amplifier channel. (See Chart 6-1.)

Delay-line action is shown in Figure 6-2. With signal delay, the leading edge of a high-speed pulse is visible on the oscilloscope screen. If a delay line is not included in the vertical channel, the leading edge of the pulse is not displayed and remains invisible off-screen. Oscilloscopes also are classified in terms of vertical-amplifier frequency response. Thus, an audio-frequency oscilloscope has a typical frequency response from 20 Hz to 20 kHz. Again, a TV service-type oscilloscope has a frequency response from

Figure 6-1. A modern solid-state triggered-sweep oscilloscope. (*Courtesy,* B & K Precision, Div. of Dynascan Corp.)

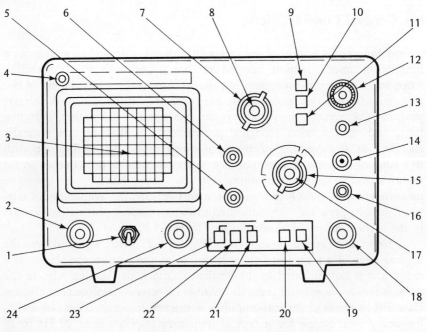

Chart 6-1.

1. POWER ON toggle switch. Applies power to oscilloscope.
2. INTENSITY control. Adjusts brightness of trace.
3. Scale. Provides calibration marks for voltage and time measurements.
4. Pilot lamp. Lights when power is applied to oscilloscope.
5. ◀▶POSITION control. Rotation adjusts horizontal position of trace. Push-pull switch selects 5X magnification when pulled out; normal when pushed in.
6. ▼ POSITION control. Rotation adjusts vertical position of trace.
7. VOLTS/DIV switch. Vertical attenuator. Coarse adjustment of vertical sensitivity. Vertical sensitivity is calibrated in 11 steps from 0.01 to 20 volts per division when VARIABLE 8 is set to the CAL position.
8. VARIABLE control. Vertical attenuator adjustment. Fine control of vertical sensitivity. In the extreme clockwise (CAL) position, the vertical attenuator is calibrated.
9. AC vertical input selector switch. When this button is pushed in the dc component of the input signal is eliminated.
10. GND vertical input selector switch. When this button is pushed in the input signal path is opened and the vertical amplifier input is grounded. This provides a zero-signal base line, the position of which can be used as a reference when performing dc measurements.
11. DC vertical input selector switch. When this button is pushed in the ac and dc components of the input signal are applied to vertical amplifier.
12. V INPUT jack. Vertical input.
13. ⏚ terminal. Chassis ground.
14. CAL⊓jack. Provides calibrated 0.8 V p-p square wave output at the line frequency for calibration of the vertical amplifier.
15. SWEEP TIME/DIV switch. Horizontal coarse sweep time selector. Selects calibrated sweep times of 0.5 μ SEC/DIV to 0.5 SEC/DIV in 19 steps when VAR/HOR GAIN control 17 is set to CAL. Selects proper sweep time for television composite video waveforms in TVH (television horizontal) and TVV (television vertical) positions. Disables internal sweep generator and displays external horizontal input in EXT position.
16. EXT SYNC/HOR jack. Input terminal for external sync or external horizontal input.
17. VAR/HOR GAIN control. Fine sweep time adjustment (horizontal gain adjustment when SWEEP TIME/DIV switch 15 is in EXT position). In the extreme clockwise position (CAL) the sweep time is calibrated.
18. TRIG LEVEL control. Sync level adjustment determines point on waveform slope where sweep starts. In fully counterclockwise (AUTO) position, sweep is automatically synchronized to the average level of the waveform.
19. TRIGGERING SLOPE switch. Selects sync polarity (+), button pushed in, or (−), button out.
20. TRIGGERING SOURCE switch. When the button is pushed in, INT, the waveform being observed is used as the sync trigger. When the button is out, EXT, the signal applied to the EXT SYNC/HOR jack 16 is used as the sync trigger.
21. TVV SYNC switch. When button is pushed in the scope syncs on the vertical component of composite video.
22. TVH SYNC switch. When button is pushed in the scope syncs on the horizontal component of composite video.
23. NOR SYNC switch. When button is pushed in the scope syncs on a portion of the input waveform. Normal mode of operation.
24. FOCUS control. Adjusts sharpness of trace.

Chart 6-1. (Cont.)

110 OSCILLOSCOPE TESTS AND MEASUREMENTS

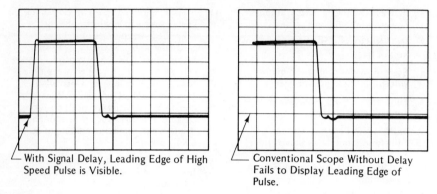

└─ With Signal Delay, Leading Edge of High Speed Pulse is Visible.

└─ Conventional Scope Without Delay Fails to Display Leading Edge of Pulse.

Figure 6-2. A vertical-amplifier delay line permits display of the leading edge in a fast-rise waveform.

approximately 20 Hz to 5 MHz. A general-purpose laboratory-type oscilloscope has a frequency response from 0 Hz (dc) to 15 MHz. An oscilloscope designed for fast-pulse measurements has high-frequency response up to 18 gigahertz (GHz).

Oscilloscopes also are classified in accordance with their low-frequency capability. Most high-performance oscilloscopes employ direct-coupled vertical and horizontal amplifiers. These instruments respond to low frequencies down to and including dc. Economy-type oscilloscopes often utilize RC-coupled amplifiers and seldom have useful response below 20 Hz. If an oscilloscope is to be used to check amplitude linearity of audio amplifiers, both its vertical and horizontal amplifiers must have high-fidelity characteristics. In other words, the oscilloscope amplifiers must have better linearity than the audio amplifiers under test. Otherwise, deficiencies in oscilloscope response would be falsely charged to the amplifier under test. Oscilloscopes that are used to check the square-wave response of audio or video amplifiers must have better transient response than the amplifier under test. Chart 6-2 summarizes basic oscilloscope information capabilities.

6-2 Rise-time Measurement and Transient Response

One of the basic specifications of a transient waveform such as a pulse or a square wave is its *rise time*. As shown in Figure 6-3, rise time is defined as the time interval between the 10 and 90 percent of maximum amplitude points along the leading edge of the wavefront. Rise time is measured with the aid of a triggered time base with calibrated sweeps. Each horizontal interval on the screen corresponds to a predetermined period, such as 1 ms (millisecond), 1 μs (microsecond), or 1 ns (nanosecond). Similarly, the fall time

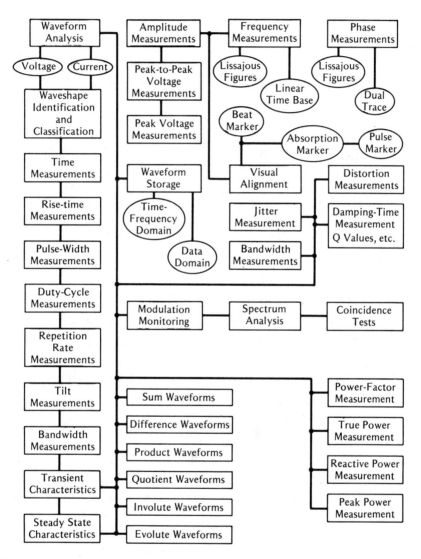

Chart 6-2.

111

112 OSCILLOSCOPE TESTS AND MEASUREMENTS

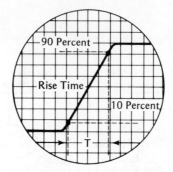

Figure 6-3. Rise time T is measured from 10 to 90 percent of maximum waveform amplitude.

(decay time) of a transient waveform is measured between the 90 and the 10 percent of maximum amplitude points along the trailing edge of the waveform. Frequency response of a vertical amplifier (or any amplifier) and rise time are related by the approximate equation:

$$T = \frac{1}{3f_c}$$

where T = rise time of the amplifier

f_c = -3 dB cutoff frequency of the amplifier

Consistent units must be used in the foregoing equation. In other words, if the frequency is stated in Hz, T will be given in seconds; if f_c is stated in kilohertz, T will be given in milliseconds; if f_c is stated in MHz, T will be given in microseconds.

6-3 Circuit Loading by Oscilloscope

Just as incorrect measurements result from objectionable circuit loading by an ac voltmeter, so are incorrect measurements obtained as a consequence of excessive circuit loading by an oscilloscope. An oscilloscope should be used with a direct input (coaxial) cable only when checking low-impedance and low-frequency circuits. High-impedance and high-frequency circuitry can be checked properly only when the oscilloscope is used with a low-capacitance probe. A typical low-capacitance probe incurs a signal-voltage loss of 90 percent and, in turn, increases the input impedance of the oscilloscope by a factor of 10 times. The amplitude loss in the probe is recovered by reserve gain provided in the vertical amplifier.

A low-capacitance probe consists basically of a series (isolating) resistor shunted by a suitable value of trimmer capacitance. When the time-constant

6-4 PULSE DISPLAY ON HIGH-SPEED SWEEP

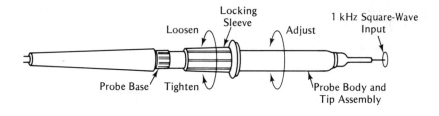

(a)

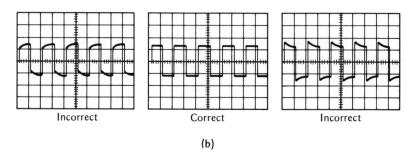

(b)

Figure 6-4. Compensation of low-capacitance probe: (a) adjustment of probe response; (b) correct and incorrect 1-kHz displays. (*Courtesy, Hewlett-Packard*)

of the probe is correctly adjusted (by adjustment of the trimmer capacitor), the probe will be free from transient distortion. This is an essential consideration in all applications. In Figure 6-4, the trimmer capacitor in the probe is adjusted by rotating the probe body with respect to the probe base so that a 1-kHz square-wave input is displayed without distortion on the screen. The probe body is then secured in position to the base by the locking sleeve. If the trimmer capacitance is too great, the reproduced waveform appears more or less integrated. On the other hand, if the trimmer capacitance is too small, the reproduced waveform appears more or less differentiated.

6-4 Pulse Display on High-Speed Sweep

A fast-rise pulse is displayed on high-speed sweep with a triggered and calibrated time base (see Figure 6-5). This pulse has a rise time of approximately 20 ns and a width of 20 μs. When displayed at a sweep speed of 0.02 ms/cm, the pulse appears to rise and fall instantaneously. In other words, the finer detail of the waveform is invisible. As the sweep speed is increased

114 OSCILLOSCOPE TESTS AND MEASUREMENTS

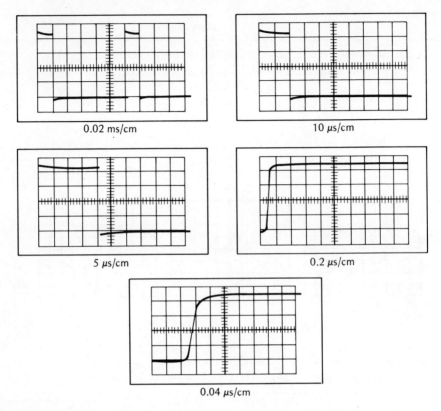

Figure 6-5. Expansion of a 0.02-msec pulse as the time-base speed is progressively increased.

to 0.2 µs/cm, it becomes apparent that the rise is not instantaneous, and that the corners of the pulse are somewhat rounded. The effect of the vertical-channel delay line also becomes evident at this sweep speed; the leading edge of the pulse is displaced from the left-hand limit of the screen area. Finally, at a sweep speed of 0.04 µs/cm, it becomes evident that the rise time of the pulse is about 0.02 µs. This typical progression is sometimes puzzling to the inexperienced operator, because this kind of waveform appears to have extremely fast rise and sharp corners on slow-speed sweep. Thus, the oscilloscope is in effect an extension of the operator's vision, and high-speed sweep action reveals waveform detail that would otherwise remain invisible.

Consider next the basic pulse-waveform characteristics shown in Figure 6-6. The high-frequency information is contained along and in the vicinity of the leading edge. The low-frequency information is contained along the top (and bottom) of the waveform. Characteristics of interest to

6-4 PULSE DISPLAY ON HIGH-SPEED SWEEP 115

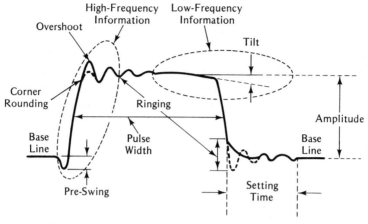

Figure 6-6. Pulse waveform characteristics.

the operator are the base-line level, amplitude, pulse width, rise time, fall time, overshoot, ringing, settling time, and tilt. The amplitude is measured in peak-to-peak volts. Nearly all oscilloscopes are designed with calibrated vertical amplifiers, so that the peak-to-peak voltage of a waveform can be easily measured. Pulse width is measured between the 50 percent of maximum amplitude points on the leading and trailing edges; it is stated in ms, μs, or ns. Fall time is measured in the same general manner as explained above for rise-time measurement.

Overshoot is stated in percentage. It is defined as the amplitude of the first maximum excursion of a pulse beyond its 100 percent amplitude level,

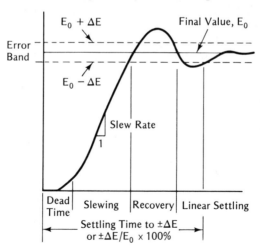

Figure 6-7. Detail of settling-time sequence.

expressed as a percentage of this reference amplitude. A ringing interval in a waveform has a particular frequency which can be measured on a calibrated sweep display. Settling time is defined as the time interval required from the start of a ringing sequence for the oscillation to enter and to remain within a specified narrow band centered on the final amplitude of the waveform. Referring to Figure 6-7, it is seen that settling time is associated with a stipulated error band, with a "dead time," and with a slew rate that is defined as the rise time of the pulse waveform. Slewing is followed by a recovery interval, and finally by a linear settling interval. Slew rate is defined as the maximum rate of change in an output voltage in response to a sudden change of input voltage.

Tilt is defined as shown in Figure 6-8. The tilt in a square waveform (or in any waveform that is normally flat-topped) is related to the low-frequency response of the circuit under test by the approximate equation:

$$f_c = \frac{2f(E_2 - E_1)}{3(E_2 + E_1)}$$

where f_c is the -3 dB cutoff frequency of the circuit (low-frequency end)

f is the square-wave frequency

E_2 and E_1 are the amplitudes indicated in Figure 6-8.

When a pulse waveform is analyzed for tilt, the factor f in the foregoing equation should be multiplied by a fraction equal to the pulse width divided by twice the repetition rate of the pulse waveform. Thus, the rise time and tilt in a reproduced square wave (or equivalent waveform) are related to the high-frequency cutoff and the low-frequency cutoff points on the circuit's frequency-response curve.

Consider next the measurement of transit time on a delay line, as illustrated in Figure 6-9. A pulse generator energizes the line through a series resistor $R1$. This resistor should have a value equal to the input impedance of the delay line, minus the impedance of generator cable

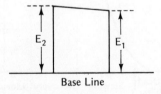

$$\text{Percent Tilt} = \frac{E_2 - E_1}{E_2} \times 100$$

Figure 6-8. Tilt in top of pulse waveform.

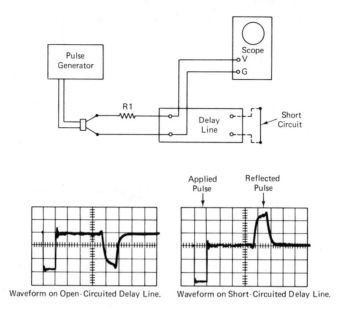

Figure 6-9. Reflection-time measurements for a pulsed delay line.

termination. Its function is to prevent any re-reflections on the line, which could confuse the display. The far end of the line may be either short-circuited or open-circuited. If it is short-circuited, the reflected pulse has opposite polarity from that of the applied pulse. If the line is open-circuited, the reflected pulse has the same polarity as that of the applied pulse. Transit time (delay time) is equal to one-half of the elapsed time from the leading edge of the applied pulse to the leading edge of the reflected pulse. Elapsed time is measured from the 50 percent of maximum amplitude points on the leading edges of the pulses.

6-5 Measurement of Time Constant

A basic time-constant measurement is made as depicted in Figure 6-10. The time constant is defined as the period required for an exponential quantity to change by an amount equal to 0.632 times the total change that will occur. In an RC circuit (Figure 6-10) the time constant is the number of seconds (or the fraction of a second) that is required for the capacitor to acquire 63.2 percent of its full voltage after a step function is applied to the RC circuit. The time constant of a capacitor that has a capacitance value C (expressed in farads) connected in series with a resistance value R (expressed in ohms) is equal to $R \times C$, and the product is in seconds. Similarly, in an RL circuit, the number of seconds required for the current to attain 63.2

118 OSCILLOSCOPE TESTS AND MEASUREMENTS

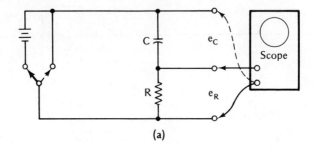

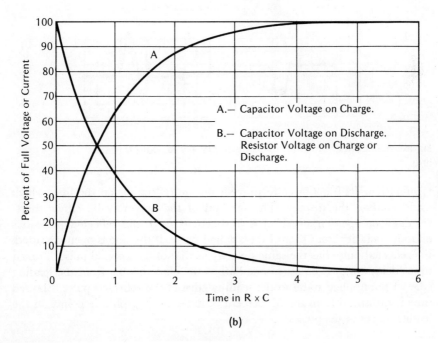

Figure 6-10. Rising and falling exponential curves: (a) basic circuit arrangement; (b) universal time-constant chart.

percent of its final value is defined as the time constant of the circuit. The time constant of an inductor that has an inductance value L (expressed in henrys) connected in series with a resistance value R (expressed in ohms) is equal to L/R, and the quotient is in seconds.

With reference to Figure 6-10, the time constant of the RC circuit is measured with a triggered-sweep oscilloscope that has a calibrated time base. The number of seconds (or the fraction of a second) required for the displayed curve to rise to 63.2 percent of its final value, or to fall to 36.8 percent of the zero level, is noted. In practice, the battery and switch shown in the diagram are replaced by a step-function generator or its equivalent.

For example, a square-wave generator is often utilized. A step function is defined as a signal that is characterized by instantaneous change. It exists in theory, but cannot be realized in practice. A square waveform that has a much faster rise than that of the circuit under test serves the purpose adequately.

6-6 Measurement of Inductor Q Value

The approximate Q value of a coil can be measured at a chosen frequency with the test setup depicted in Figure 6-11. Capacitor $C1$ is selected to resonate the inductor at a desired frequency. A small pulse voltage is coupled into the circuit under test by means of a "gimmick" that consists of one or two turns of insulated wire wound around the inductor lead. In turn, a ringing waveform is displayed on the oscilloscope screen. This waveform is a combination of a sine wave and an exponential wave. Note that the sinusoidal component is not a true sine wave, because it is decaying in amplitude. From an analytical viewpoint, it represents a sine wave that has been amplitude-modulated by an exponential wave. In turn, the fundamental sine wave has become accompanied by a spectrum of many other

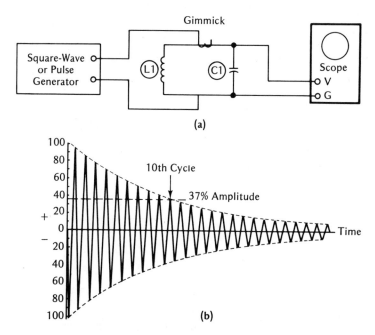

Figure 6-11. An exponentially damped sine waveform: (a) test setup; (b) ringing waveform indicates approximate circuit Q value.

120 OSCILLOSCOPE TESTS AND MEASUREMENTS

sideband frequencies. As indicated in the diagram, a ringing waveform provides a ready, if approximate, measure of the circuit Q value. From a practical viewpoint, this Q value is given by X_L/R, where R denotes the ac resistance of the ringing circuit. In the example given, the ringing waveform decays to 37 percent amplitude over an interval of 10 cycles. The Q value is given approximately by the product of the number of cycles within this interval and pi (3.14 +). Therefore, the approximate Q value of the ringing circuit in this example is 31.4. Again, suppose that 15 cycles were required in another situation for decay of the ringing waveform to 37 percent amplitude. In such a case, the Q value would be approximately 47.1.

When the operator needs to make a highly accurate measurement of the Q value from a ringing waveform, the chart shown in Figure 6-12 should be employed. As indicated in the diagram, the percentage ratio in amplitude between successive peaks is measured on the oscilloscope screen. In turn, the Q value of the ringing circuit is given by the graph. For example, suppose that the second cycle in a ringing waveform has 90 percent amplitude with respect to the first cycle. The chart shows that the corresponding Q value is 31. Note in passing that the third cycle will have 90 percent amplitude with

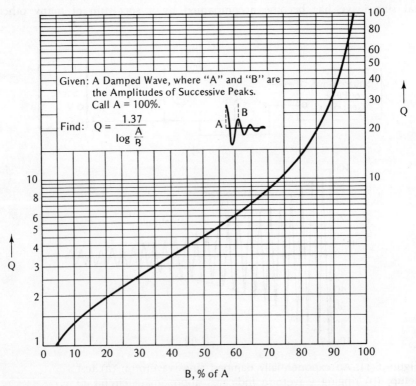

Figure 6-12. Q value of a damped resonant system in terms of successive peak amplitudes. (*Courtesy*, Hewlett-Packard)

respect to the second cycle in the foregoing example. Similarly, the fourth cycle will have 90 percent amplitude with respect to the third cycle. This decay characteristic defines an exponentially damped sine wave.

6-7 Phase Measurement

Several methods are employed for phase measurement with an oscilloscope. A basic method that can be used with any type of oscilloscope is depicted in Figure 6-13. Observe that the time base is synchronized externally from the

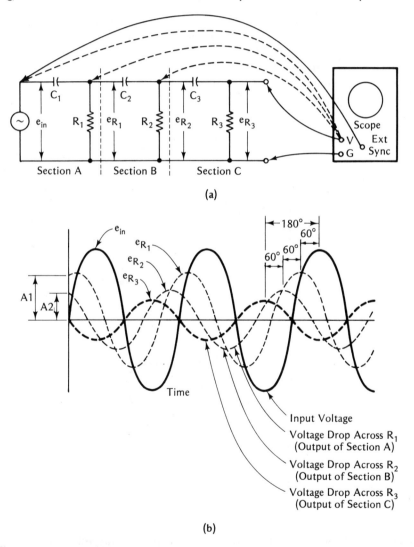

Figure 6-13. Waveforms in an L-type three-section phase shifter: (a) Test setup; (b) progressive waveforms show phase relations.

122 OSCILLOSCOPE TESTS AND MEASUREMENTS

input voltage. As the vertical-input test lead is moved progressively through the circuit under test, each waveform starts in a phase that is referenced to the input voltage. This method indicates whether the phase angle of the waveform under test leads or lags the input phase; it also shows the relative amplitudes of the progressive waveforms. To measure phase angles precisely, the operator should note the initial amplitude of the waveform, such as $A1$, and the peak amplitude of the waveform. Then, the ratio of initial amplitude to peak amplitude is calculated, and the corresponding angle is found from a table of sines in terms of the arcsin ($\sin^{-1}$) value.

Another basic method of phase-angle measurement is shown in Figure 6-14. Here, the input and output voltages from a 2-port unit or network are applied to the horizontal and vertical amplifiers of an oscilloscope to produce a Lissajous pattern. The operator centers the pattern precisely on the screen, and then measures the intervals M and N, as indicated in Figure 6-14(b). Finally, the ratio M/N is calculated and the value of the phase

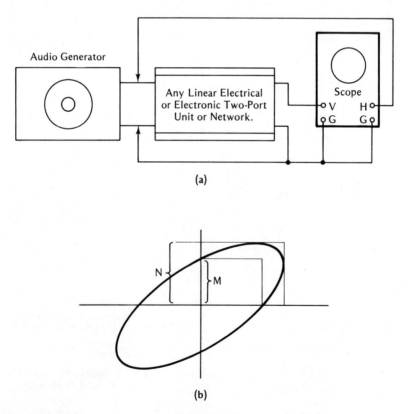

Figure 6-14. Phase measurement with Lissajous pattern: (a) test setup; (b) phase angle is equal to arcsin M/N.

angle is determined from a table of sines. Otherwise stated, the ratio M/N is equal to the sine of the phase angle; the phase angle is equal to the arcsin or $\sin^{-1}$ of this ratio. It is essential that the audio generator have a good sine waveform. Accurate results also require good amplitude linearity in the oscilloscope's horizontal and vertical amplifiers.

Still another method of phase-angle measurement is shown in Figure 6-15. This technique requires a dual-trace oscilloscope. The Channel-A input lead is applied at one point in the circuit, and the Channel-B input lead is applied at another point in the circuit. Accordingly, two patterns are

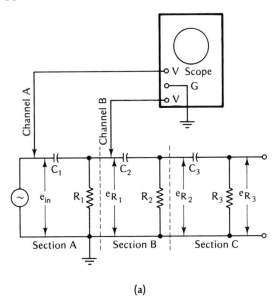

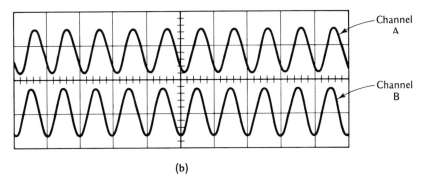

Figure 6-15. Phase measurement with a dual-trace oscilloscope: (a) test setup; (b) dual traces display a phase shift.

124 OSCILLOSCOPE TESTS AND MEASUREMENTS

displayed on the oscilloscope screen. In the example of Figure 6-15(b), the Channel-B trace is displaced somewhat to the left with respect to the Channel-A trace. This displacement defines the amount of phase shift from Section A to Section B. The relative displacement also shows that the Channel-B trace is leading the Channel-A trace. One of the advantages of this dual-trace technique is that the waveforms are displayed simultaneously, so that evaluation is facilitated.

6-8 Linearity Measurement

Waveform linearity, as for a ramp or sawtooth, is easily measured by means of a conventional time-base display—triggered and calibrated sweeps are not required. The only requirement for the oscilloscope is that it must have highly linear vertical and horizontal amplifiers. Otherwise, deficiencies in the instrument amplifiers would be falsely charged to the equipment under test. In Figure 6-16, the presence of nonlinearity is determined by holding a straight-edge along the ramp interval of the waveform. Then, if nonlinearity is found, the operator notes the amount by which the ramp departs from its true amplitude. This value is then divided by the theoretically correct amplitude of the waveform, and is expressed as a percentage. In the example shown, the interval of nonlinearity corresponds to a 16 percent error.

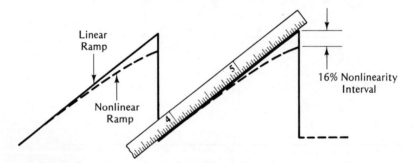

Figure 6-16. An example of 16 percent nonlinearity in a sawtooth waveform.

6-9 High Frequency Measurements

Various kinds of high-frequency measurements are made with the oscilloscope. Among these, visual-alignment measurements are most common. A basic test setup is depicted in Figure 6-17 for checking the frequency response of a television front end (tuner). In normal operation, a display is obtained as shown in (b). If a malfunction, such as regeneration, should occur, a distorted display is produced, as shown in (c). It is necessary, but

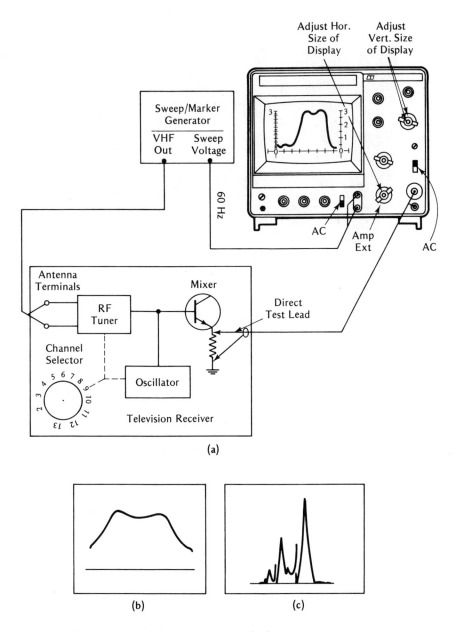

Figure 6-17. Example of a front-end visual alignment arrangement: (a) test setup; (b) normal RF-tuner frequency-response curve; (c) regenerative frequency-response curve; (d) appearance of a sweep-and-marker generator; (e) RF response curve limits.

126 OSCILLOSCOPE TESTS AND MEASUREMENTS

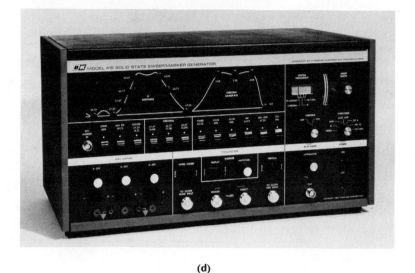

(d)

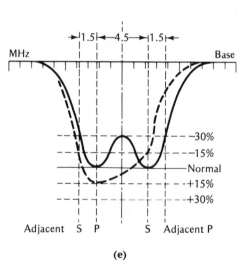

(e)

Figure 6-17. (Cont.)

not sufficient, that a properly shaped pattern appear on the oscilloscope screen. In other words, the center frequency of the pattern must be correct. In addition, the video-carrier and sound-carrier frequencies must fall at the top of the response curve, as seen in Figure 6-18. This determination is made with the aid of a marker generator. This instrument places a *pip* on the curve at the frequency to which the marker generator is set. In case the marker frequencies do not appear at correct locations on the curve, the operator must adjust the tuning slugs in the resonant circuitry to obtain the specified response.

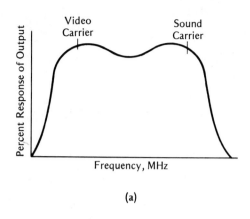

(a)

Frequency Reference Chart

Channel Number	Picture Carrier Frequency (MHz)	Sound Carrier Frequency (MHz)	Receiver VHF Oscillator Frequency (MHz)
2	55.25	59.75	101
3	61.25	65.75	107
4	67.25	71.75	113
5	77.25	81.75	123
6	83.25	87.75	129
7	175.25	179.75	221
8	181.25	185.75	227
9	187.25	191.75	233
10	193.25	197.75	239
11	199.25	203.75	245
12	205.25	209.75	251
13	211.25	215.75	257

(b)

Figure 6-18. RF tuner frequency-response curve: (a) curve shape with marker placement; (b) channel operating frequencies. (*Courtesy,* B & K Precision, Div. of Dynascan Corp.)

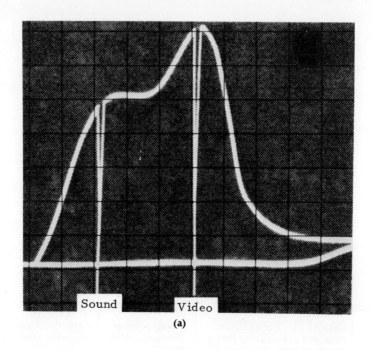

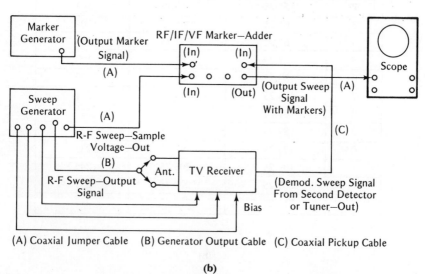

Figure 6-19. Post-injection marking of RF frequency-response curve: (a) example of large post-injection markers; (b) typical test setup; (c) appearance of a post-marker/sweep generator. (*Courtesy*, Heath Co.)

6-9 HIGH FREQUENCY MEASUREMENTS 129

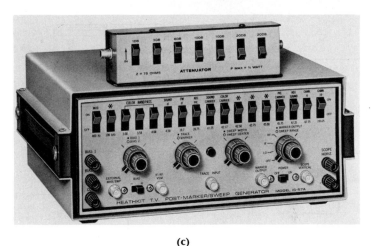

Figure 6-19. (Cont.)

Each channel of a television front end should be checked for correct frequency response—picture and sound carrier frequencies are noted in Figure 6-18(b). Oscillator frequencies are generally adjusted in another procedure, after VHF alignment has been completed. There has been a trend to the use of post-injection marking of response-curve frequencies. Post injection has the advantage that the marker signal does not pass through the tuned circuits under test. Therefore, very large markers can be utilized, if desired, without any overloading effect on the tuner circuitry. As an illustration, very large post-injection sound and video carrier markers are illustrated in Figure 6-19. A sample of the sweep-frequency voltage is mixed with the marker-generator signal in a separate marker-adder unit.

Next, the operator checks the IF response, as pictured in Figure 6-20. This procedure involves test frequencies in the 45-MHz range. Although the oscilloscope cannot respond to this high frequency, the sweep signal becomes demodulated through the video detector. In turn, the displayed frequency-response curve contains frequencies of 60, 120, 180, etc. Hz. Similarly, when a front end is under test, the high-frequency sweep signal becomes demodulated through the mixer transistor before it is applied to the oscilloscope. If the test frequency exceeds the response capability of the oscilloscope, a demodulator must always be included in a high-frequency test setup. If a demodulator is not available in the section under test, it must be provided externally. For example, a frequency-response check of an individual IF stage is made with the aid of a demodulator probe connected to the oscilloscope. The probe functions to demodulate the sweep and marker signal so that its wave envelope is developed. This envelope contains low frequencies to which any oscilloscope can respond.

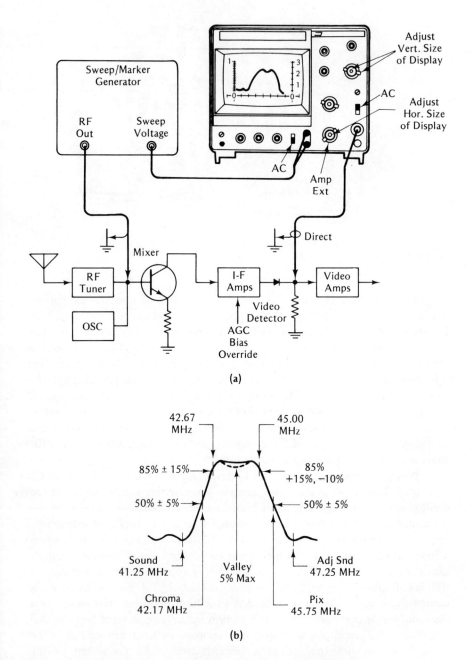

Figure 6-20. Example of IF alignment procedure: (a) test setup; (b) IF response curve, with permissible tolerances noted; (c) frequency relations along the IF response curve.

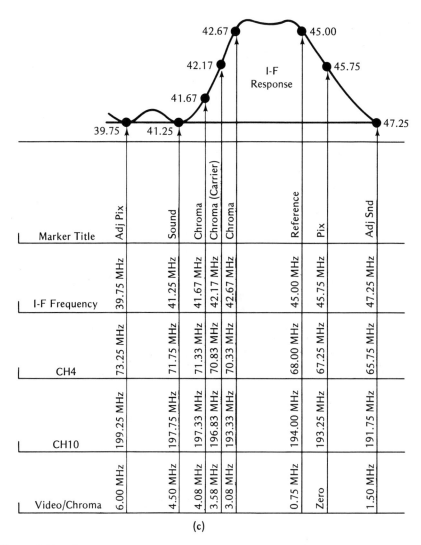

Figure 6-20. (Cont.)

132 OSCILLOSCOPE TESTS AND MEASUREMENTS

6-10 Transistor Measurements

An oscilloscope is used to make transistor measurements with the aid of a semiconductor curve tracer, as illustrated in Figure 6-21. The instrument operates by applying a certain bias voltage to the transistor under test, and then sweeps the collector voltage through some preset range (from zero to 10 volts). This varying collector voltage is applied to the horizontal deflecting

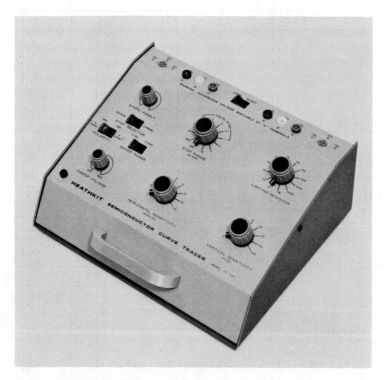

Figure 6-21. A semiconductor curve tracer. (*Courtesy,* Heath Co.)

plates of the CRT in an oscilloscope and moves the spot horizontally. At the same time, the collector current that flows through the transistor is converted into a corresponding voltage and is applied to the vertical deflecting plates. The result is a display of one curve for the collector family of the transistor as depicted in Figure 6-22. If two more steps of a staircase bias voltage are applied to the transistor, a display of three collector-family curves results, as

6-10 TRANSISTOR MEASUREMENTS 133

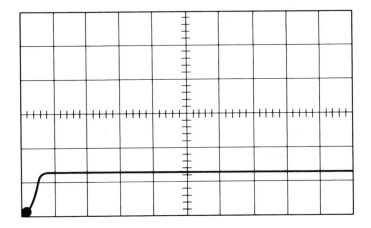

Figure 6-22. Initial display of a collector-family curve.

shown in Figure 6-23. A larger number of curves will be displayed if the instrument is set for a greater number of steps in the staircase bias waveform.

Consider next an example of output-impedance measurement shown in Figure 6-24. The output impedance (collector resistance) of a transistor is defined as a change in collector voltage, divided by the corresponding change in collector current. In this example, the operator takes an interval of 25 V in the horizontal direction; the corresponding current change in the vertical direction is 0.8 mA. Accordingly, the output impedance of the transistor is equal to 31,250 Ω. Note that this is a dynamic or incremental impedance. The measured value will depend to some extent upon the particular voltage interval and the particular curve chosen by the operator,

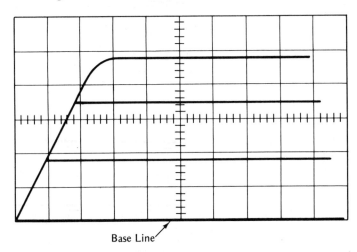

Figure 6-23. Display of three collector-family curves by successive changes in base-bias voltage.

134 OSCILLOSCOPE TESTS AND MEASUREMENTS

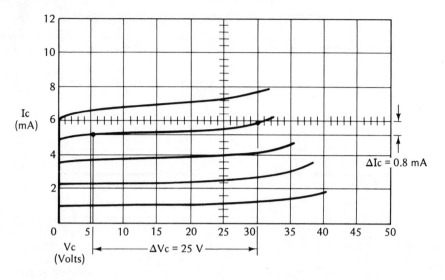

Figure 6-24. Example of output-impedance measurement.

because a transistor is not an entirely linear device. In general, the region for measurement is chosen in the vicinity of the normal operating point for the transistor.

Transistor beta (current gain) is measured as illustrated in Figure 6-25. A collector potential of 4 V is stipulated in this case. The two curves chosen for the measurement correspond to base currents of 0.05 mA and 0.15 mA,

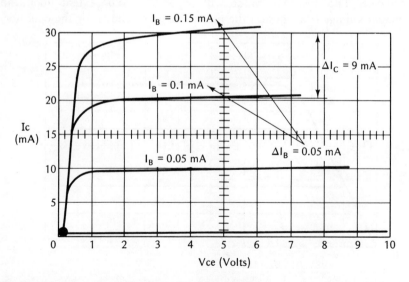

Figure 6-25. Measurement of transistor beta value.

respectively. Corresponding collector-current values are 21 mA and 30 mA. Otherwise stated, a change of 0.05 mA in base current produces a change of 9 mA in collector current. Therefore, the beta value in this region is equal to 9/0.05, or 180. This means that the current gain of the transistor is 180 times. A curve tracer is highly informative, because it displays the complete collector family for a transistor and minimizes the time and labor that is required to check out a suspected defective transistor.

6-11 Television Network Tests and Measurements

To check for errors in the transient response of television networks, a vertical interval test signal (VITS) is included in the vertical synchronizing interval of color-TV waveforms (see Figure 6-26). This VITS waveform includes a

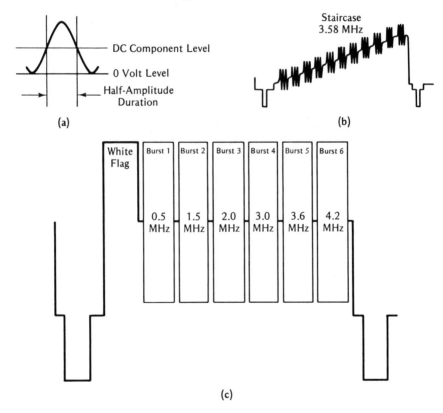

Figure 6-26. Vertical interval test signal waveforms: (a) $\sin^2$ pulse; (b) staircase/3.58-MHz waveform; (c) multiburst signal; (d) appearance of encoded VITS waveform in the vertical synchronizing interval.

136 OSCILLOSCOPE TESTS AND MEASUREMENTS

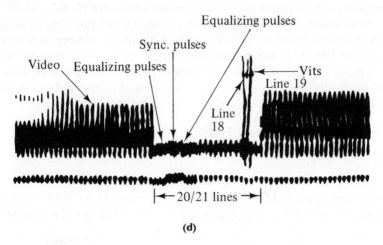

(d)

Figure 6-26. (Cont.)

$\sin^2$ pulse, a staircase/3.58-MHz waveform, and a multiburst signal. The $\sin^2$ pulse is particularly useful for checking the transient response of a television network, because its waveshape is a good approximation of a picture element. A $\sin^2$ waveform is produced when a sine wave is multiplied by itself. This product is a double-frequency sine wave with a dc component, as shown in Figure 6-27. In other words, a $\sin^2$ waveform has a positive excursion only, whereas a sine waveform has both positive and negative excursions. The dc component in a $\sin^2$ waveform has a value equal to the peak value of either of the original sine waves (multiplier or mutiplicand). The width of a $\sin^2$ pulse is called its half-amplitude duration (h.a.d.). This width is generally stated in T units, where T is equal to the duration of one picture element in the TV network. In a 4-MHz video system, T is equal to 0.125 μsec.

Since a $\sin^2 T$ pulse (or simply T pulse) has a wave shape that is practically the same as that of a video signal corresponding to a picture element, it is a good indicator of transient response in a TV network. Although the $\sin^2$ pulse appears highly compressed when the video signal is displayed on 30- or 60-Hz sweep, it becomes readily visible when displayed on high-speed sweep [see Figure 6-26(a)]. A TV network also may be tested for transient response with a $2T$ pulse. In a 4-MHz system, a $2T$ pulse will have a half-amplitude duration of 0.25 μsec. A $2T$ pulse has fewer high-frequency harmonics than does a T pulse, and it is used to check the system transient response at somewhat lower frequencies. Sometimes a test also is made with a $20T$ pulse. In a 4-MHz system, a $20T$ pulse has a half-amplitude duration of 2.5 μsec. This $20T$ pulse is used to amplitude-modulate the 3.58-MHz color subcarrier in a color-TV system. Its purpose is to check the difference in gain that may be present between the low-frequency and the

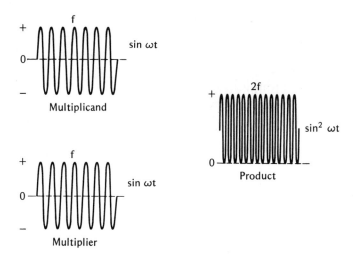

Figure 6-27. A sine-squared waveform is produced by multiplication of a sine wave by itself.

high-frequency ends of the video-frequency spectrum. It also is used to check the phase relation (delay time) between signals at the high and at the low ends of the video-frequency spectrum. These checks are made by observing the amount and kind of distortion that is displayed by the reproduced pulse after it has passed through the system.

As shown in Figure 6-28, the $\sin^2$ pulse is followed by a window pulse in the VITS waveform. This window pulse provides the operator with a peak-to-peak voltage reference for low-frequency signals and indicates whether the high-frequency pulse may have become attenuated in passage through the system. In addition, the window pulse will indicate the presence of low-frequency linear distortion in terms of undershoot, tilt, or overshoot. Undershoot and overshoot are related to phase irregularities; tilt is related to impaired low-frequency response.

The transient response in a vestigial-sideband system differs from that in a double-sideband system, so that predistortion of the video signal is required at the transmitter in order to achieve optimum system response. This response is pictured in Figure 6-29. When a step response (such as a window pulse) is applied to a 4-MHz vestigial sideband system (VSB system), the reproduced step response at the receiver will be distorted. It will exhibit undershoot, slow rise, smear (tilt), and ringing. After introduction of phase equalization and amplitude compensation at the transmitter, the reproduced step response at the receiver becomes optimized, although it cannot be made distortionless. This optimized response has equal ringing intervals at its black and white levels, plus a comparatively fast rise, with no smear interval.

138 OSCILLOSCOPE TESTS AND MEASUREMENTS

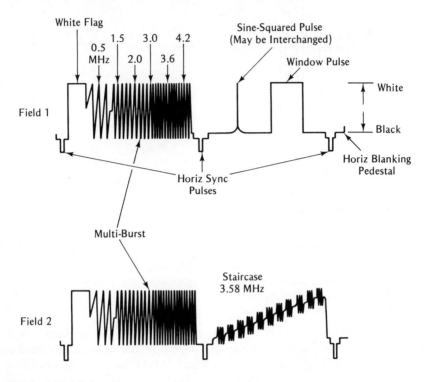

Figure 6-28. A complete VITS waveform.

Next, consider a video network check with a T pulse (Figure 6-30). If a T pulse is passed through a system that has incorrect frequency response, but correct phase response, a typical form of distortion is displayed by the reproduced pulse. This distortion consists of lobes preceding and following the main excursion of the reproduced T pulse. These lobes consist of undershoots and ringing excursions. Because the leading lobes and the trailing lobes are the same in the example of Figure 6-30(b), the indication is that the system phase response is correct and only the frequency response is incorrect. Note that if the system phase response and its frequency response are both correct, a reproduced T pulse will still show a first-lobe development. The reason for this is that the frequency response of a T pulse extends past the 4-MHz cutoff point. Thus, it is the second-lobe development in Figure 6-30(b) that indicates incorrect frequency response of the system.

Consider next the reproduction of a $2T$ pulse in a TV network. A $2T$ pulse in a 4-MHz system has a half-amplitude duration of 0.25 μsec. The basic usefulness of the $2T$ pulse is that its frequency spectrum does not extend past the 4-MHz cutoff point in the transmission channel. If the system has normal frequency response, the $2T$ pulse is reproduced in

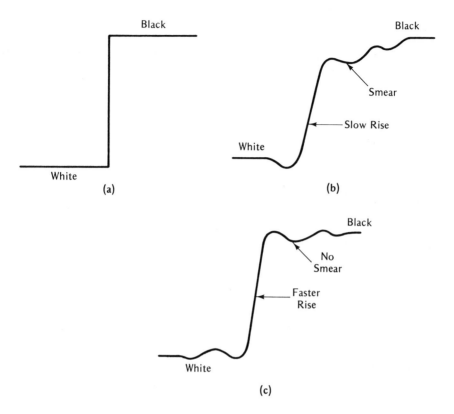

Figure 6-29. Step response in a VSB 4-MHz system: (a) ideal step voltage; (b) system response without predistortion; (c) system response with predistortion.

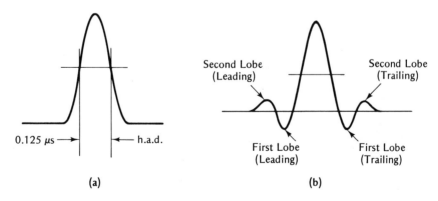

Figure 6-30. Basic T-pulse ($\sin^2$ pulse) response, 4-MHz system: (a) input T pulse; (b) output T pulse showing incorrect frequency response and correct phase response.

undistorted form without any first-lobe development. Note that a $2T$ pulse test is subject to the limitation that it cannot clearly indicate any abnormalities toward the high-frequency end of the transmission channel. It will indicate abnormalities up to approximately 75 percent of the channel cutoff frequency, and it provides no useful information concerning the final 25 percent of the high-channel frequency response. Therefore, it is good practice to check video-channel response with both a T pulse and with a $2T$ pulse.

Consider next the indication of system phase abnormalities by a T pulse. When the system phase characteristic is nonlinear, the reproduced T pulse is not symmetrical. Instead, the pattern becomes skewed, either toward the leading edge or toward the lagging edge. In addition, as depicted in Figure 6-31, an unsymmetrical lobe development occurs. In other words, lobes at the left-hand end of the pattern indicate high-frequency lead in the system, while lobes at the right-hand end of the pattern indicate high-frequency lag in the system. These distortions are corrected by means of phase equalizers. Phase equalization does not eliminate lobe development. It does, however, serve to make the reproduced T pulse symmetrical, to reduce

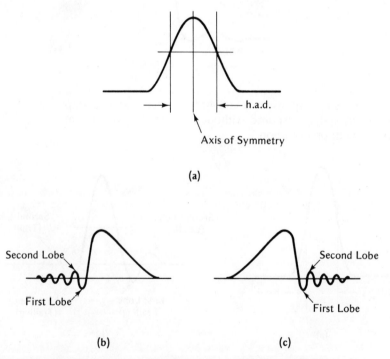

Figure 6-31. Nonlinear phase indication by a T pulse: (a) normal T pulse display; (b) system high frequencies leading; (c) system high frequencies lagging.

the lobe amplitude, and to distribute the lobe development equally at the left- and right-hand ends of the pattern.

High-frequency attenuation in the system causes a T pulse to reproduce at subnormal amplitude, and it also increases the width of the reproduced pulse. If the high-frequency attenuation is gradual, these effects on the T pulse shape are not accompanied by ringing (lobe development). On the other hand, a rapid attenuation of the high frequencies causes considerable ringing, with only a minor reduction in pulse amplitude. The ringing is determined by the frequency at which the frequency-response curve drops substantially. As an illustration, in a conventional frequency-response curve with rapid attenuation at its high-frequency end, the ringing frequency is determined by the cutoff frequency (-3dB cutoff point). It is not a common occurrence for a video system to develop a rising high-frequency response (high-frequency hump). However, if this condition happens to occur, the ringing frequency will be determined by the hump frequency in the response curve.

Television network tests also are made to determine differential gain and differential phase. Differential gain is defined as the difference between the ratio of the output amplitudes of a small high-frequency sine-wave signal measured at two stipulated dc-component levels, and unity. Otherwise stated, this difference is the value of the differential gain under the stated test conditions. Differential phase is defined as the difference in phase shift through a TV system for a small high-frequency sine-wave signal at two stipulated dc-component levels. It follows that if the transfer characteristic of a system is nonlinear, both differential-gain and differential-phase distortion will occur. Differential gain is usually measured with a staircase signal, as

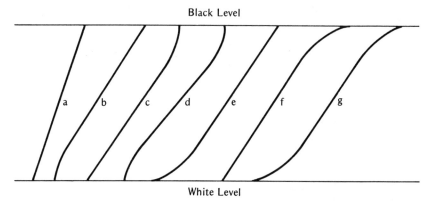

Figure 6-32. Basic transfer characteristics: (a) linear; (b) white stretch; (c) black stretch; (d) white and black stretch; (e) white compression; (f) black compression; (g) white and black compression.

provided in a VITS waveform. Differential phase is generally measured with a vectorscope or equivalent phase-indicating instrument.

Differential gain is undesirable in a video signal because the gray range becomes distorted. If the transfer characteristic is linear, as depicted in Figure 6-32(a), the gray range will be reproduced correctly. However, if white compression is present, the lighter grays are reproduced as whites. Again, if black stretch occurs, the darker grays are reproduced as black. If both white stretch and black stretch are included in the transfer characteristic, only medium grays are reproduced correctly; lighter grays are reproduced as white and darker grays are reproduced as black. In the event that white compression is present, the lighter grays are reproduced as white. If black compression occurs, the darker grays are reproduced as black. When both white compression and black compression are included in the transfer characteristic, darker grays are reproduced as black and lighter grays are reproduced as white; only medium grays are reproduced correctly. The distorting effect of black compression on a staircase signal is pictured in Figure 6-33.

Nonlinear transfer characteristics are compensated at the terminal equipment by passing the video signal through a nonlinear amplifier that is adjusted to provide an opposing curvature in its transfer characteristic.

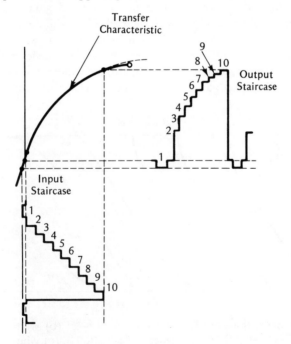

Figure 6-33. Distorting effect of black compression on a staircase signal.

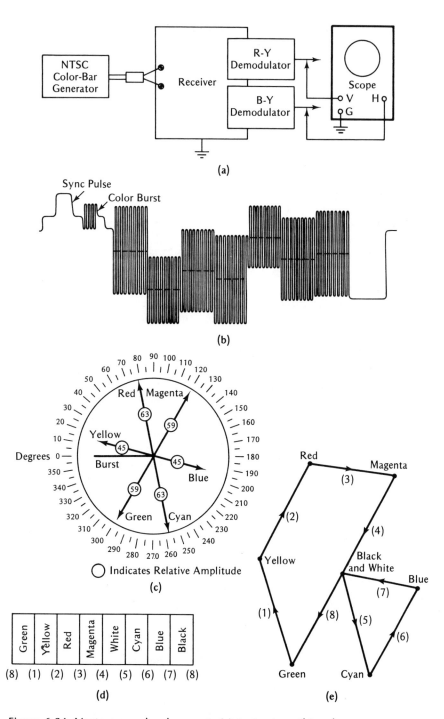

Figure 6-34. Vectorgram development: (a) test setup; (b) color-bar signal; (c) phases of primary and complementary hues; (d) color-bar display on picture-tube screen; (e) vectorgram displayed by vectorscope.

144 OSCILLOSCOPE TESTS AND MEASUREMENTS

Correct compensation is indicated by the transmission of an undistorted staircase signal. After the transfer characteristic has been linearized, the resulting video signal is checked for linearity of its phase characteristic. Differential phase distortion is particularly undesirable in a color-TV signal because it makes hue dependent upon the brightness level. In other words, if the lighting level changes in a color scene that has differential phase distortion, the hues will appear to shift accordingly.

Phase measurement with a vectorgram pattern is pictured in Figure 6-34. The basic test setup is depicted in (a); an NTSC (National Television Systems Committee) signal shown in (b) is demodulated on the $R-Y$ and $B-Y$ axes and the outputs are applied to the vertical and horizontal channels, respectively, of an oscilloscope. This is called a vectorscope arrangement. All television broadcast systems employ some form of NTSC color-bar signal for test purposes. This signal comprises the primary and the complementary colors, with corresponding phases shown in (c). The waveform depicted in (b) produces the color-bar pattern shown in (d) on the screen of a color picture tube. When the demodulator outputs in (a) are applied to a scope as indicated, a vectorgram is displayed on the CRT screen as diagrammed in (e). Progressive spot travel is indicated by the consecutive numbers.

If a keyed-rainbow signal is utilized, as seen in Figure 6-35, another type of vectorgram is developed. A keyed-rainbow signal, also termed an offset color subcarrier, a sidelock signal, or a linear phase sweep, is characterized by successsive bursts that differ in phase by 30 degrees. This relationship is indicated in (b). In theory, a keyed-rainbow signal would produce a vectorgram of the form depicted in Figure 6-36(a). In practice, however, receiver circuit characteristics often result in a somewhat unsymmetrical pattern, as shown in (b). Most TV service technicians utilize a keyed-rainbow signal, whereas color-TV engineers and broadcast technicians use an NTSC signal.

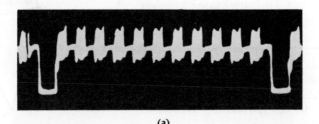

(a)

Figure 6-35. Keyed-rainbow color-bar signal characteristics: (a) typical oscilloscope display; (b) color-bar phase and hue identifications; (c) appearance of a keyed-rainbow color-bar generator. (*Courtesy*, Heath Co.)

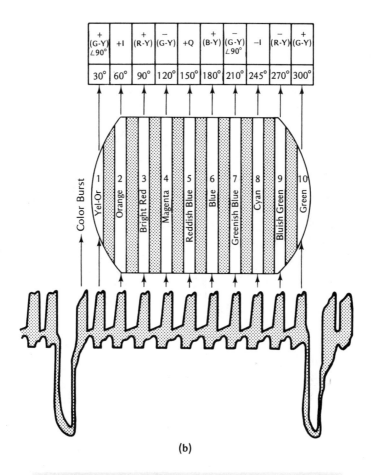

(c)

Figure 6-35. (Cont.)

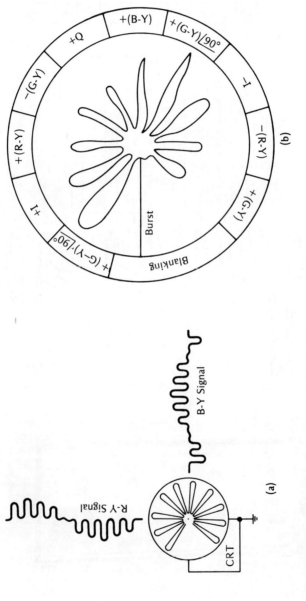

Figure 6-36. Keyed-rainbow vectorgrams: (a) ideal; (b) typical.

146

6-12 TV Receiver Waveform Characteristics

It is instructive to observe the characteristics of TV receiver waveforms in various circuit conditions. As an illustration, the horizontal sync pulse is one of the key waveforms in the video channel. A sync pulse normally progresses from the input to the output of the video channel without a large change in wave shape. On the other hand, when a circuit malfunction is present, the wave shape of the sync pulse becomes changed substantially. In Figure 6-37, the waveform from a video signal generator, for example, provides a sync pulse with relatively fast rise and square corners. If the video waveform is processed by wide-band circuitry, its wave shape will remain essentially unchanged. On the other hand, if the video waveform is processed by circuitry with subnormal bandwidth, the rise time of the sync pulse will be increased and the corners of the pulse will become rounded, as seen in the illustration. This form of distortion should not be confused with clipping distortion that results from passage through an overdriven amplifier. For example, Figure 6-38 shows a sync pulse with normal proportions and the

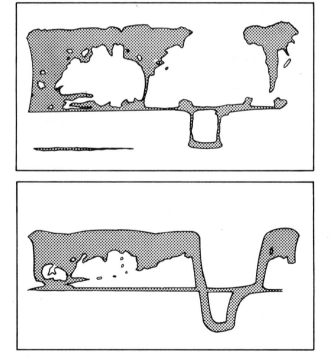

Figure 6-37. Sync pulse becomes distorted by amplifier with inadequate bandwidth.

148 OSCILLOSCOPE TESTS AND MEASUREMENTS

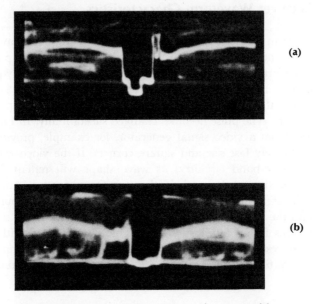

Figure 6-38. Sync pulse becomes clipped through an overdriven amplifier.

same pulse with a clipped sync tip after passage through a peak-clipping amplifier. This form of distortion results in loss of horizontal sync action.

It should not be supposed that a sync pulse necessarily maintains correct amplitude levels at various points through a video channel. Circuit malfunctions can attenuate the video signal. Some kinds of malfunction cause the sync pulse to have an abnormally high level. As a rough rule of thumb, it is considered that a video signal is within acceptable amplitude limits if the sync pulse does not vary more than ±20 percent from its specified peak-to-peak voltage. An oscilloscope screen display of a sync pulse with normal amplitude, +20 percent amplitude, and −20 percent amplitude is depicted in Figure 6-39. Just as the factors involved in dc voltage tolerances are highly complex, so are the factors associated with sync-pulse amplitude tolerances.

Various other types of pulses are encountered in TV receiver circuitry. For example, AGC pulses, blanking pulses, AFC pulses, and horizontal driving pulses occur in a normally operating circuit. In abnormal operation of intercarrier-sound circuits, sync-buzz pulses are encountered. A blanking pulse, for example, is characterized by amplitude, width, and shape. These features are illustrated in Figure 6-40. The normal (bogie) amplitude of a blanking pulse is generally considered to have a permissible tolerance of ±20 percent. In addition, the pulse has a specified (bogie) width that has a comparatively tight tolerance. If the pulse is too wide, part of the picture

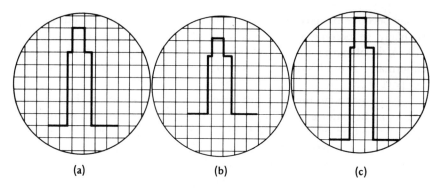

Figure 6-39. Example of ±20 percent tolerance: (a) specified amplitude; (b) −20 percent amplitude; (c) +20 percent amplitude.

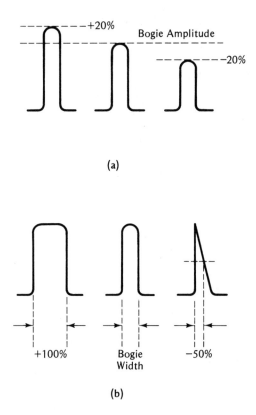

Figure 6-40. Examples of amplitude and width variations: (a) ±20 percent amplitude variations: (b) +100 percent and −50 percent width variations.

150 AC VOLTAGE MEASUREMENTS

display becomes blanked out. If the pulse is too narrow, vertical-retrace lines are likely to become visible over part of the screen. As seen in Figure 6-40(b), a blanking pulse may become misshapen, so that its width is subnormal. Again, this defect is likely to result in visible vertical-retrace lines over part of the screen.

Next, consider the forms of sync-buzz pulses shown in Figure 6-41. Sync buzz becomes evident as a harsh 60-Hz buzzing sound, particularly when the picture display has a light background. It is caused by malfunctions in the intercarrier sound channel that result in amplitude-modulation of the intercarrier sound signal by the picture signal. When this occurs, the

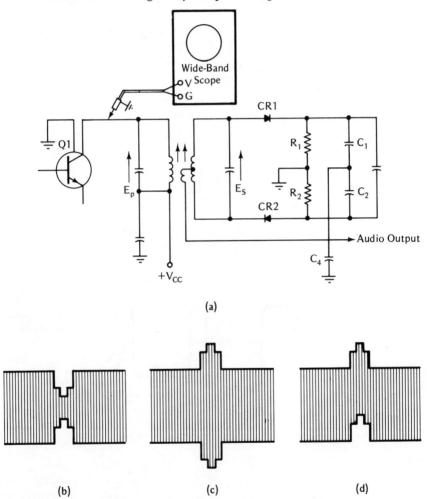

Figure 6-41. Three types of sync-buzz waveforms: (a) test setup; (b) downward modulation; (c) upward modulation; (d) linear mix.

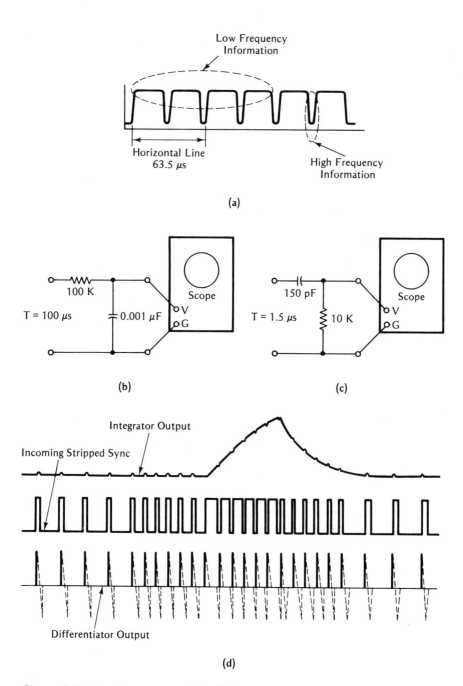

Figure 6-42. Low-frequency and high-frequency information in the sync train: (a) regions of low- and high-frequency information in the vertical-sync pulse; (b) differentiating circuit separates high-frequency component; (c) integrating circuit separates low-frequency component; (d) waveforms at inputs and at outputs of differentiating and integrating circuits.

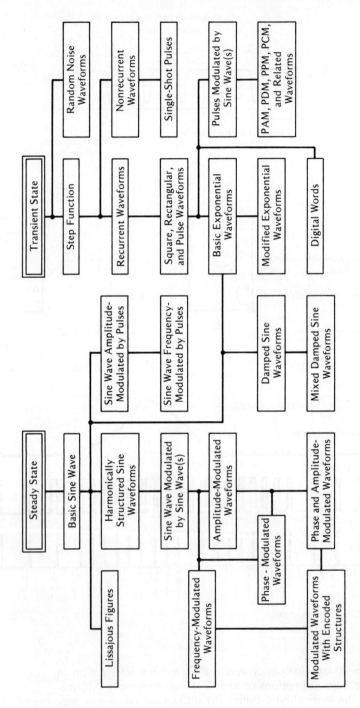

Chart 6-3.

vertical sync pulses introduce a 60-Hz interference signal with which the ratio detector may be unable to contend. A check for the presence of sync-buzz pulses is made as shown in the diagram. Note that downward modulation, as depicted in Figure 6-41(a), is the most difficult type of interference with which a ratio detector must contend. To correct this malfunction, the system operation must be approximately linearized throughout.

Key sync-channel waveforms are shown in Figure 6-42. Observe that the flat tops in a vertical sync pulse contain low-frequency information, whereas the serrations contain high-frequency information. In turn, vertical-frequency timing pulses are developed by an integrating circuit, whereas horizontal-frequency timing pulses are developed by a differentiating circuit. Stripped (clipped) sync waveforms are applied to the inputs of the integrating and differentiating circuits. As shown in the diagram, the integrating circuit normally develops a broad pulse with a 60-Hz repetition rate, whereas the differentiating circuit normally develops a narrow pulse with a 15,750-Hz repetition rate. Observe that this repetition rate increases to 31,500 Hz for the duration of the serrated interval in the vertical sync pulse. Basic waveform relations are summarized in Chart 6-3.

6-13 Ignition Waveform Analysis

The main parts of a standard ignition system are the battery, the coil, the distributor, and the spark plugs. Located within the distributor (Figure 6-43) is a cam that revolves to open and close the breaker points. While the breaker points are closed, a complete circuit is formed to permit battery current to flow through the primary winding of the coil. This current causes an intense magnetic field to form around the primary winding of the coil during the dwell time. When the breaker points open, the magnetic field collapses rapidly and induces a high voltage in the secondary winding which can easily approach 15 to 20 kilovolts. This surge of high voltage, which occurs each time that the breaker points open, is fed back to the distributor where the rotor (a rotating switch) applies it to the proper spark plug in the firing order. The condenser across the distributor breaker points provides the clean electrical break necessary to produce proper high voltage and prevents the points from arcing and burning.

Complex voltage signal pulses are produced by the various components of the ignition system. An ignition analyzer (Figure 6-44) converts these electrical pulses into a visual pattern that is displayed on the CRT screen. Comparison of the actual displayed pattern with the normal pattern that is produced by a properly operating ignition system enables the operator to detect any deviation from normal operation, and to pinpoint the trouble

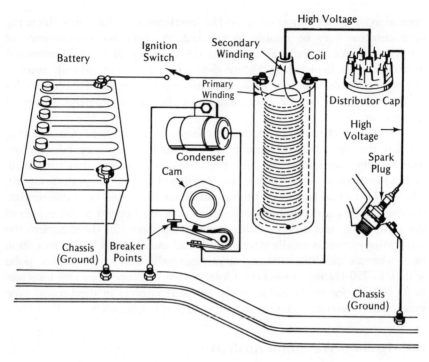

Figure 6-43. Basic automotive ignition system. (*Courtesy,* Heath Co.)

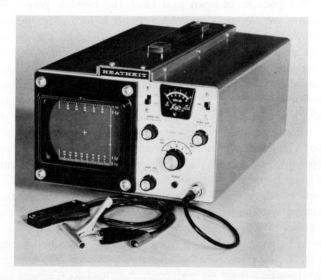

Figure 6-44. An ignition analyzer is a specialized form of oscilloscope. (*Courtesy,* Heath Co.)

6-13 IGNITION WAVEFORM ANALYSIS 155

area. Consequently, the operator needs to know how each part in an ignition system affects a normal waveform display.

An ignition analyzer will display a superimposed primary or a superimposed secondary pattern in which the firing patterns of all engine cylinders are shown simultaneously one on top of the other. It will also display a primary or a secondary "parade" pattern in which the firing patterns of all engine cylinders are shown from left to right across the screen in their normal firing order. The primary superimposed pattern is most useful for locating troubles that may occur owing to a poor connection anywhere between the vehicle battery and the grounded side of the breaker points in the distributor. The secondary superimposed pattern is most useful in locating troubles that may occur in the high-voltage circuits between the ignition coil and the spark plugs. The "parade" pattern is used to determine if one or more of the firing patterns are not normal, and, if so, which engine cylinders are involved.

Consider the primary waveform display depicted in Figure 6-45. The system relations may be summarized as follows:

A (Points open signal)	1.	Breaker points open. High voltage produced in coil's secondary winding; spark plug fires.
A-B (Spark zone)	1.	High voltage is directed by the distributor to the correct spark plug for firing.
B-C (Coil-condenser zone)	1.	Spark plug stops firing. Coil-condenser oscillations show unused coil energy being dissipated to ground.
C (Points close signal)	1.	Start of current flow through coil primary.
C-D (Dwell zone)	1.	Current through the coil primary rebuilds the magnetic field around both windings.

This completes the firing cycle for one cylinder. At D, the breaker points open again and the firing-pattern sequence is repeated for the next cylinder in the firing order.

Consider next the secondary waveform display depicted in Figure 6-46.

156 OSCILLOSCOPE TESTS AND MEASUREMENTS

The system relations may be summarized as follows:

A (Plug firing signal)	1.	Breaker points open. Produces high voltage in coil's secondary winding; spark plug fires.
A-B (Spark zone)	1.	Firing time of spark plug. Once that the spark is started at A, a lower voltage sustains the firing to B.
	2.	Because the coil-condenser oscillations that occur in the primary circuit are not reflected, the secondary pattern shows a horizontal line during the time that the spark plug fires. This line, called the "spark line", is very informative, since any deviations in this zone reflect difficulties in the high-voltage circuits.
B-C (Coil-condenser zone)	1.	Spark plug stops firing. Coil-condenser oscillations show unused coil energy being dissipated to ground.
C (Points close signal)	1.	Start of current flow through primary coil.
	2.	When breaker points close, there is a voltage induced in the secondary winding which oscillates for a short period of time. This signal, which is shown in the secondary pattern, is very important, since it reflects the proper closing of the points.
C-D (Dwell zone)	1.	Current flow through the coil primary rebuilds the magnetic field around both windings.

The ignition analyzer illustrated in Figure 6-44 is connected into an

6-13 IGNITION WAVEFORM ANALYSIS   157

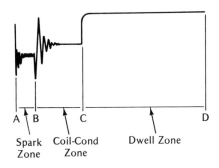

Figure 6-45. Normal primary waveform. (*Courtesy,* Heath Co.)

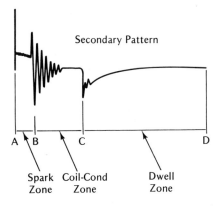

Figure 6-46. Normal secondary waveform. (*Courtesy,* Heath Co.)

ignition system as shown in Figure 6-47. A normal parade pattern and a normal superimposed pattern are shown in Figure 6-48. Distorted waveforms and associated ignition malfunctions are summarized in Figure 6-49.

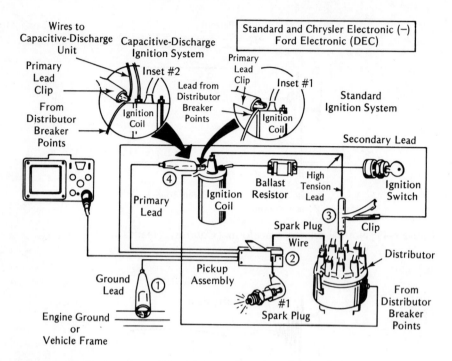

Figure 6-47. Connection of ignition analyzer into ignition system. (*Courtesy*, Heath Co.)

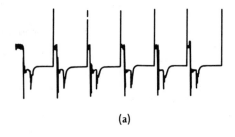

(a)

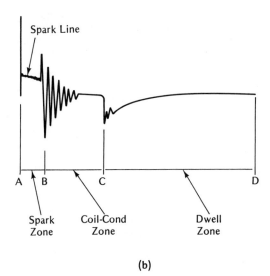

(b)

Figure 6-48. Ignition waveform patterns: (a) normal parade pattern; (b) normal superimposed pattern. (*Courtesy,* Heath Co.)

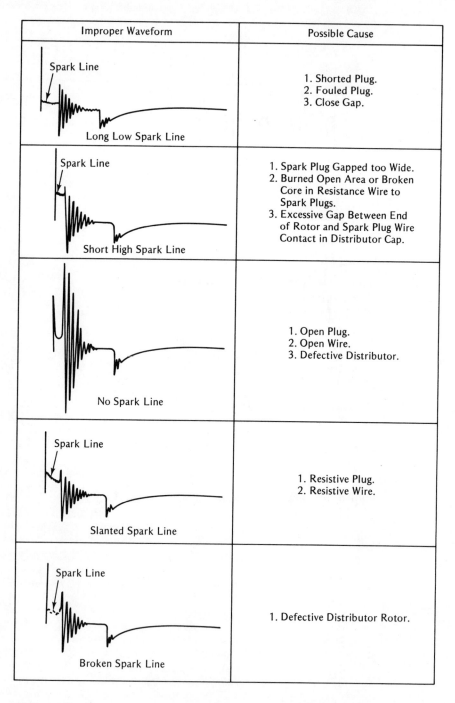

Figure 6-49. Summary of distorted waveforms and associated ignition malfunctions. (*Courtesy,* Heath Co.)

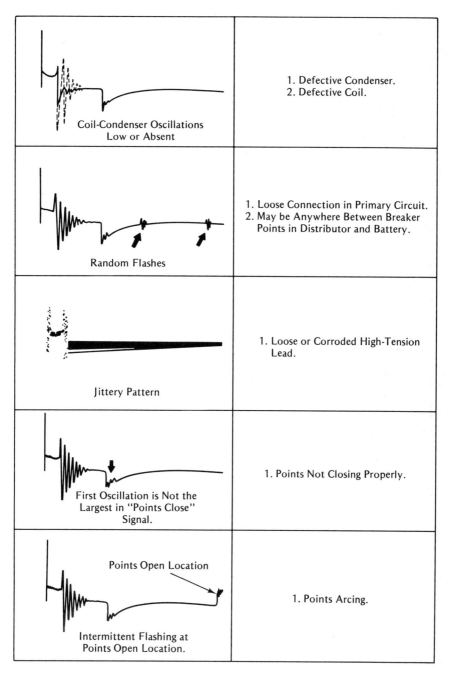

Figure 6-49. (Cont.)

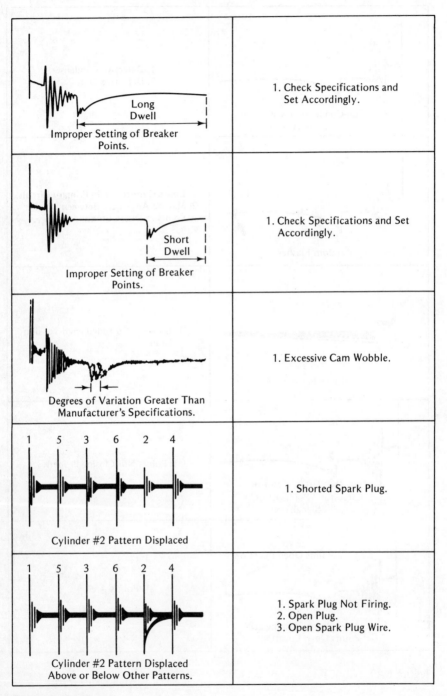

Figure 6-49. (Cont.)

Review Questions

1. How may an oscilloscope be compared with a voltmeter?
2. Explain the meaning of *rise time*.
3. What is the function of a low-capacitance probe?
4. Describe the relation between slew rate and rise time.
5. How is the time constant of an RC circuit defined?
6. Discuss the measurement of delay time on a line.
7. How are phase measurements made with an oscilloscope?
8. Explain how a visual alignment procedure is accomplished.
9. What is the function of a transistor curve tracer?
10. Why is a vertical interval test signal utilized?
11. Describe a $\sin^2$ waveform.
12. How is a T pulse pattern interpreted?
13. Define a transfer characteristic.
14. How is a staircase signal pattern evaluated?
15. Discuss the development of a vectorgram.

chapter seven

Audio Measurements

7-1 General Considerations

A wide range of electrical and electronic measurements is made in audio technology. These include measurements of frequency response, voltage gain, harmonic distortion, intermodulation distortion, transient response, power output, power bandwidth, noise output, input impedance, output impedance, input-port frequency characteristics, and music-power capability. Although harmonic distortion and intermodulation distortion are in the same test category, both measurements are often made. Transient response is usually checked with a 2-kHz square-wave signal. This is technically a test, rather than a measurement, but it is very informative in the evaluation of amplifier performance. Power-bandwidth measurement is more meaningful than power-output measurement, because the power-bandwidth value relates to the useful high-fidelity power capability of an amplifier. A measurement of music-power capability is made with a high-level pulse signal; it is indicative of the amplifier's ability to process brief high-amplitude peak signals in musical programs. It is often possible to make an audio test or measurement in more than one way. One method may be simpler than another; again, one method may be more accurate than another. One method can serve as a useful crosscheck of another. For example, an intermodulation distortion measurement provides a generalized crosscheck of

166 AUDIO MEASUREMENTS

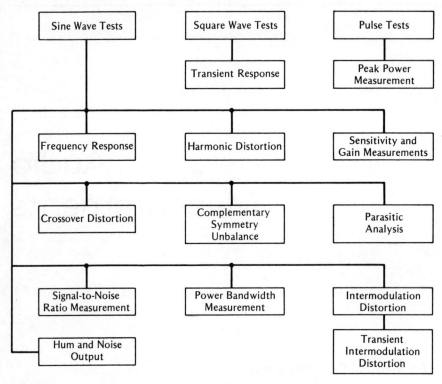

Chart 7-1.

a harmonic distortion measurement. In some cases, a choice of method will be determined by the operator's equipment availability. An overview of basic audio-frequency tests and measurements is given in Chart 7-1.

7-2 Measurement of Frequency Response

Frequency response measurement is made with the amplifier under test energized by a sine-wave signal. An audio generator such as illustrated in Figure 7-1 is utilized as a signal source. It is helpful to use a generator that has an output-level meter, so that the input signal can easily be maintained at a constant amplitude as the signal frequency is varied. A frequency-response test setup is shown in Figure 7-2. The amplifier is terminated with its rated value of load resistance. Either an oscilloscope or an ac voltmeter can be employed as an output indicator. It is more informative to use an oscilloscope because serious distortion in the test waveform becomes immediately apparent. An initial frequency-response check is generally made at maximum rated power output from the amplifier. The power output is equal to E^2/R, where E is the rms voltage measured across the load resistor R_L at 1 kHz, and R is the ohmic value of the load resistor.

7-2 MEASUREMENT OF FREQUENCY RESPONSE 167

Figure 7-1. An audio signal generator with an output level meter and a wide-range attenuator. (*Courtesy*, Leader Electronics)

Although no industry standards have been established, it is generally considered that high-fidelity operation requires a frequency response from at least 20 Hz to 20 kHz, within ±1 dB. The reference amplitude is taken at 1 kHz. Observe in Figure 7-2 that the voltage gain of the amplifier can be measured at the same time that its frequency response is checked. In other words, the voltage gain is equal to the ratio of the output voltage to the input voltage. Note also that the anticipated voltage gain will depend upon the particular input port of the amplifier that is utilized. In the example of Figure 7-3, the amplifier has a rated sensitivity of 1.5 mV for its tape-input port. This means that an input signal of 1.5 mV will normally produce an output signal of 1 V into a load of 10 kilohms. Similarly, the phono input port of the amplifier is rated for a sensitivity of 8 mV, the tuner input port is rated for 250 mV, and the microphone input port is rated for 250 mV.

Although an audio power amplifier is designed for a flat (uniform) frequency response, a preamplifier has different rated frequency responses for each input port. Standard frequency responses for phono and tape input ports are depicted in Figure 7-4. A tuner input port normally has a flat frequency response. Note also that the frequency response of an input port will be affected by the setting of the tone control(s). Accordingly, each tone control should be set for flat response when the amplifier frequency response is checked.

Consider next how the power bandwidth of an audio amplifier is measured (refer to Figure 7-5). The amplifier is operated first at maximum rated power output and its harmonic distortion measured at 1 kHz. Then, the power output is reduced by 3 dB and a lower frequency limit is determined at which the harmonic distortion is the same as in the first test

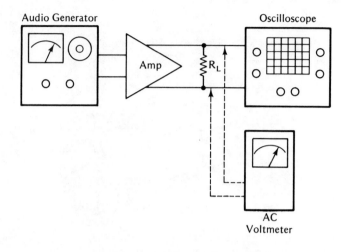

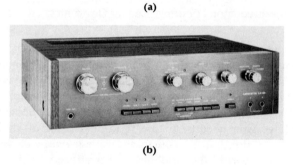

Figure 7-2. Audio amplifier frequency response: (a) test setup; (b) appearance of a 4-channel audio amplifier. (*Courtesy, Lafayette Electronics*)

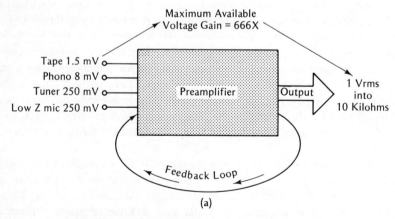

Figure 7-3. Preamplifier arrangement: (a) input/output characteristics; (b) configuration.

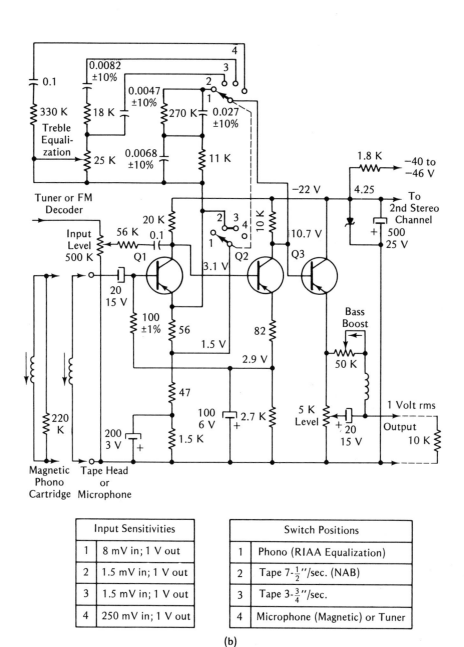

Figure 7-3. (Cont.)

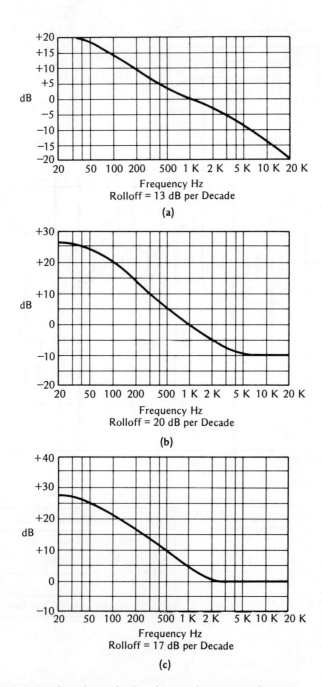

Figure 7-4. Examples of standardized input frequency characteristics: (a) RIAA equalization curve for playback of records; (b) NAB standard playback curve for 7.5 in./s tape; (c) MIRA playback curve for 3.75 in./s tape.

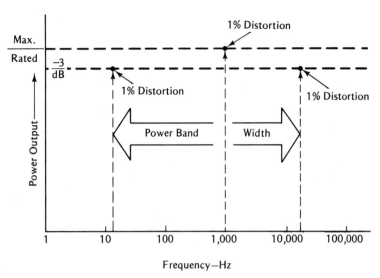

Figure 7-5. Representation of audio-amplifier power bandwidth.

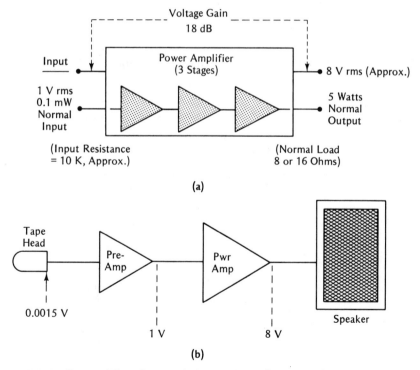

Figure 7-6. Audio-amplifier characteristics: (a) input/output voltage and power values; (b) signal-voltage levels through an audio channel.

(1 percent in this example). Next, with the power output at the same −3 dB level, a higher frequency limit is determined at which the harmonic distortion is the same as in the first test (1 percent in this example). These two limiting frequencies define the power bandwidth of the amplifier.

Consider next the amplifier characteristics illustrated in Figure 7-6. The voltage gain of the power amplifier is 18 dB. In other words, an input signal level of 1 V is amplified to a level of 8 V. This 1-V signal is developed across an input resistance of 10 kΩ, or, the input signal power in this example is 0.1 mW. At the output of the amplifier, an 8-V signal is developed across a load resistance in the range from 8 to 16 Ω. In turn, an output power level of 5 W is typical. The frequency response and power bandwidth will be less for the combination of preamp and power amp than it is for either of the amplifiers alone. This deterioration in performance, however, is small for well-designed audio amplifiers.

7-3 Measurement of Harmonic and Intermodulation Distortion

A high-fidelity amplifier is normally rated for less than 1 percent harmonic distortion at maximum rated power output. Harmonic distortion is measured with a harmonic-distortion meter (HDM), as shown in Figure 7-7. The audio generator is set to 1 kHz and the amplifier is driven to maximum

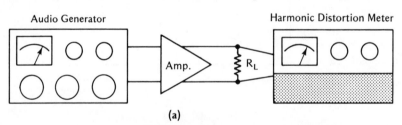

(a)

(b)

Figure 7-7. Harmonic distortion test setup: (a) equipment connections; (b) appearance of harmonic-distortion analyzer. (*Courtesy,* Heath Co.)

7-3 MEASUREMENT OF DISTORTION 173

rated power output. In turn, the percentage harmonic distortion is indicated by the HDM. A low-distortion audio generator must be used in this test. In other words, unless the generator has less distortion than the amplifier under test, the generator error will be falsely attributed to the amplifier under test. Note that a harmonic distortion meter can be used to check the percentage distortion of an audio generator by feeding the generator signal directly into the harmonic distortion meter. Operation of a harmonic distortion meter is based upon trap action. If a hi-fi amplifier developed no distortion, its output would be a pure sine wave, devoid of harmonics. In practice, however, some distortion always occurs, although it may be very small. Distortion is accompanied by the development of harmonic frequencies. When a distorted signal is processed by a harmonic distortion meter, the fundamental frequency is trapped out of the signal, and the harmonics are passed into the instrument indicator section. The amplitude of these harmonics is indicated by the meter on a percentage scale.

Since a harmonic distortion meter cannot distinguish between different types of distortion, other types of test equipment are required. Most amplifier distortion is caused by amplitude nonlinearity. In other words, the output from the amplifier is not strictly proportional to the input amplitude. This kind of distortion increases as the input signal level to the amplifier is increased. If an amplifier is overdrivem, its transfer characteristic becomes highly nonlinear. Overdrive is accompanied by compression or clipping distortion. (See Chart 7-2).) Note also that a harmonic distortion meter cannot distinguish between nonlinear (amplitude) distortion and noise. One clue that suggests the presence of noise is the indication of a higher percentage of distortion as the test-signal level is decreased. It must be emphasized, however, that this clue by itself is insufficient evidence of noise in the output signal.

Another kind of distortion that can occur in push-pull output amplifiers is crossover distortion. This is a form of nonsinusoidal output that is caused by insufficient forward bias on the push-pull output transistors. Note that a clue that suggests the presence of crossover distortion is the same as was mentioned above concerning noise output. That is, an HDM indicates a higher percentage of distortion as the test-signal level is decreased. As before, it must be emphasized that this clue by itself is insufficient evidence of crossover distortion in the output signal.

Still another type of apparent distortion that can occur in audio amplifiers is hum voltage in the output signal. This malfunction may either cause an increase or a decrease in the HDM percentage distortion indication, depending on the source of the hum voltage. However, there is a reliable test for hum that can be made with a HDM. The test frequency is set to 59 Hz and the pointer on the meter is observed for 1-Hz oscillation. If oscillation occurs, it is an indication that 60-Hz hum is present. If oscillation does not occur, however, the test is repeated at 119 Hz. Then, if oscillation occurs, it indicates that 120-Hz hum is present.

174 AUDIO MEASUREMENTS

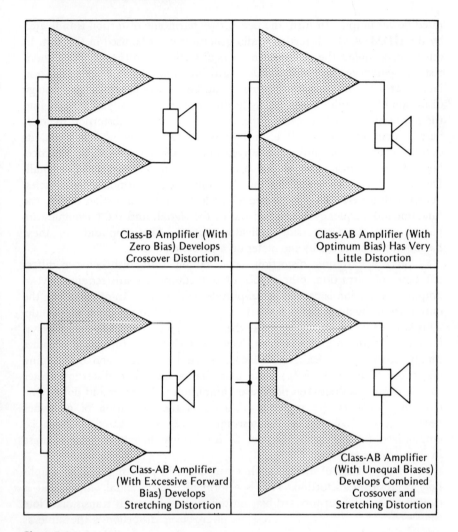

Chart 7-2.

Another method of harmonic-distortion measurement employs a spectrum analyzer, as depicted in Figure 7-8. This is a more informative instrument than the basic harmonic-distortion meter, because it dissects the waveform into its individual harmonic (or other) components and displays them along a frequency scale on the screen of the spectrum analyzer. It operates by utilizing the waveform to modulate a comparatively high-frequency oscillator. In turn, sideband frequencies are generated in correspondence to the frequency components of the modulating waveform. The oscillator frequency is swept by the deflection voltage for the cathode-ray tube (CRT). In turn, the sideband frequencies are swept, and they are passed sequentially by a high-Q bandpass filter. Finally, the filter output is

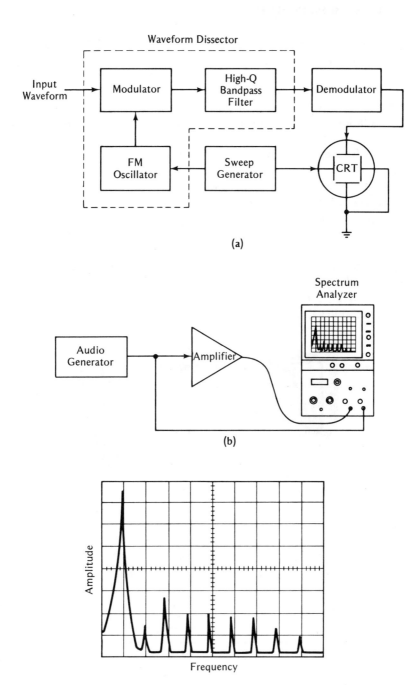

Figure 7-8. Spectrum analyzer checks harmonic distortion: (a) basic block diagram; (b) test setup; (c) typical screen display.

176 AUDIO MEASUREMENTS

demodulated and applied to the deflecting plates in a CRT. The screen pattern shows the frequency and the amplitude of each distortion component.

Intermodulation distortion is related to harmonic distortion, in that both types of distortion are caused by amplitude nonlinearity. However, intermodulation distortion involves the generation of frequencies in the output that are equal to the sums and differences of integral multiples of the component frequencies present in the input signal. An intermodulation (IM) test signal is a two-tone wave (see Figure 7-9); it is typically formed from a 60-Hz and a 6-kHz signal (both sine waves). The 60-Hz component generally has several times the amplitude of the 6-kHz component. When the two-tone signal is processed by an amplifier, any amplitude nonlinearity will cause the 6-kHz signal to be amplitude-modulated by the 60-Hz signal. An IM analyzer contains filter and demodulator sections for picking out the amplitude-modulation components. These components are applied to a meter and indicated on a percentage IM scale.

In general, intermodulation distortion and harmonic distortion will have comparable values for a particular amplifier. However, there is usually

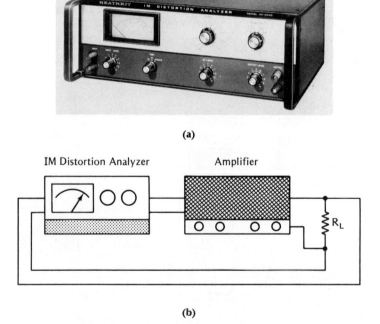

Figure 7-9. Intermodulation distortion measurement: (a) appearance of IM distortion analyzer; (b) test setup. (*Courtesy, Heath Co.*)

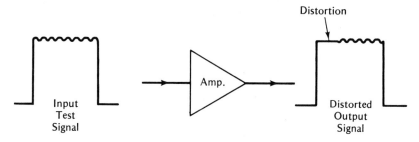

Figure 7-10. Example of transient intermodulation distortion (TIM).

some variation in values. For example, intermodulation distortion could be somewhat higher than harmonic distortion at low-test frequencies, and lower than harmonic distortion at high-test frequencies. Some variation often will be found between the percentage of harmonic distortion at a low-test frequency and at a high-test frequency. Both HD and IM percentages increase rapidly as the high-frequency cutoff point of an amplifier is approached. A similar increase is found as the low-frequency cutoff point is approached in RC-coupled amplifiers. This increase results from phase shift through the amplifier circuitry, which becomes greater in the vicinity of cutoff. The normal negative feedback action in the amplifier becomes progressively positive, thereby increasing the inherent distorting action of the amplifier, instead of decreasing it.

A form of intermodulation distortion encountered in some amplifiers that have a very large amount of negative feedback is called *transient intermodulation distortion* (TIM). An example of this type of distortion is depicted in Figure 7-10. It occurs following an abrupt transition in signal level. An amplifier is checked for transient intermodulation distortion with a test-signal waveform consisting of a square wave with a superimposed high-frequency sine wave, as shown in Figure 7-10. The output waveform is observed on an oscilloscope screen. If the amplifier is free from transient intermodulation distortion, the output waveform will be a precise replica of the input waveform. On the other hand, if TIM is present, more or less of the high-frequency sine-wave component will be absent following the leading edge of the waveform.

7-4 Measurement of Music Power Capability

It was noted previously that the music-power rating of an amplifier is checked with a pulse waveform. Vocal and musical waveforms often have narrow and high peaks, as is illustrated in Figure 7-11. In turn, an

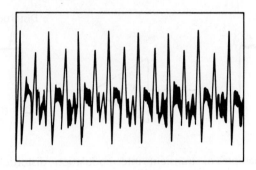

Figure 7-11. Vocal and musical waveforms have sharp high peaks.

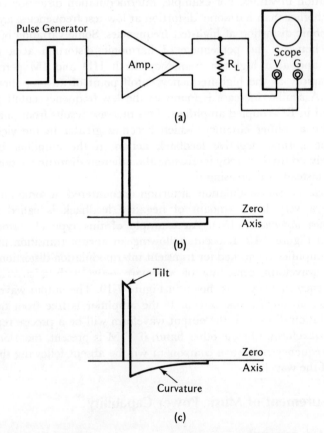

Figure 7-12. Pulse test of amplifier: (a) test setup; (b) input pulse; (c) distorted output pulse; (d) a high-quality pulse generator. (*Courtesy,* Hewlett-Packard)

(d)

Figure 7-12. (Cont.)

amplifier's capability for processing high-amplitude peak signals is of practical concern. A test pulse waveform is well adapted to this requirement. A pulse generator and oscilloscope are connected to the amplifier under test, as depicted in Figure 7-12. The test is ordinarily made with a pulse width of 1 millisecond, and with a pulse repetition rate of 100 pulses per second (pps). First, a low-amplitude pulse voltage is applied, and its amplitude is gradually increased as the operator observes the oscilloscope screen. At some point, it will be observed that the top of the pulse starts to tilt and the lower excursion of the pulse starts to curve, as seen in Figure 7-12(c). This distortion occurs because the filter capacitors in the amplifier's power supply cannot maintain a constant supply voltage for the duration of the pulse. At this point, a measurement of the pulse amplitude will show that its peak power exceeds the rated rms power output of the amplifier. This peak-power value represents the music-power capability of the amplifier.

7-5 Measurement of Impedance, Inductance, and Capacitance Values

An impedance bridge such as that illustrated in Figure 7-13 is useful for checking inductance, capacitance, and resistance values in audio circuitry. It measures impedance components, from which impedance values can be calculated. An inductance bridge is the only type of instrument that can measure inductance values directly. The exemplified bridge contains a low-Q inductance section and a high-Q section, termed a Hay bridge and a Maxwell bridge, respectively. Null indication is provided by a 100 μA

180 AUDIO MEASUREMENTS

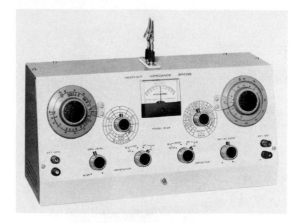

Figure 7-13. Examples of bridges: (a) an RCL (impedance) bridge; (b) a capacitor bridge. (*Courtesy*, Heath Co.)

galvanometer. (See Chart 7-3.) A Wheatstone bridge section measures resistance values from 0.1 Ω to 10 MΩ. The capacitance-bridge section measures capacitance values from 100 pF to 100 µF. Inductance values from 0.1 mH to 100 H can be measured. Q-values from 0.1 to 1000, and of dissipation factors from 0.002 to 1 are indicated. All measurements are ordinarily made with a built-in 1-kHz sine-wave generator. However, an external audio oscillator may be used to operate the bridge at a selected frequency.

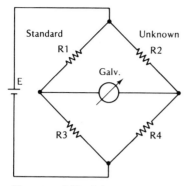

Wheatstone Bridge Balances an Unknown Resistor Against a Standard Resistor.

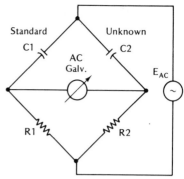

Basic Capacitance Bridge Balances an Unknown Capacitor Against a Standard Capacitor.

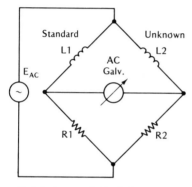

Basic Inductance Bridge Balances an Unknown Inductor Against a Standard Inductor.

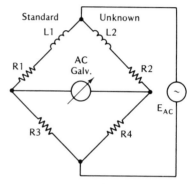

Elaborated Version of Basic Inductance Bridge Balances an Unknown Inductive Impedance Against a Standard Inductive Impedance.

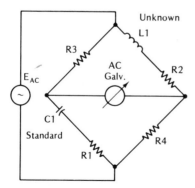

Hay Bridge Balances an Unknown Inductive Impedance Against a Standard Capacitive Impedance; Standard Capacitor and Standard Resistor are Connected in Series.
(Preferred for High-Q Inductors)

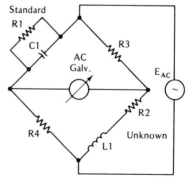

Maxwell Bridge Balances an Unknown Inductive Impedance Against a Standard Capacitive Impedance; Standard Capacitor and Standard Resistor are Connected in Parallel.
(Preferred for Low-Q Inductors)

Chart 7-3.

182 AUDIO MEASUREMENTS

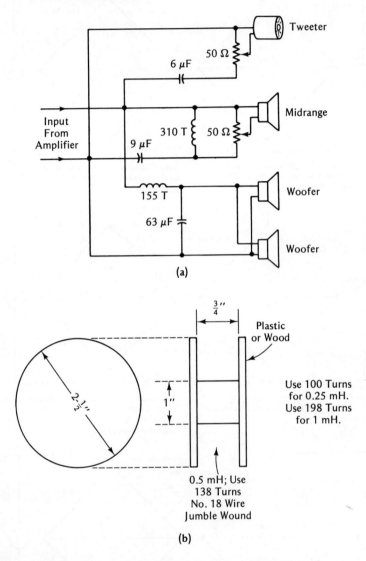

Figure 7-14. Crossover networks: (a) woofer, midrange, and tweeter crossover arrangement; (b) typical inductor construction.

An inductance bridge is needed to check the inductance values of coils used in speaker crossover networks (see Figure 7-14). Although various "rule of thumb" procedures can be used to wind inductors to approximate values, a bridge measurement is necessary to determine inductance values accurately. Similarly, a bridge is required to measure the impedance of a speaker, or of a speaker system. An impedance bridge also finds use for

7-6 Stereo Decoder Separation Measurement

measuring the inductances of chokes in power suppplies. Inductors used in tone-control networks can be checked. A bridge can be used to measure the input impedance of an amplifier. It will indicate the resistive component of the input impedance, plus the value of any inductive or capacitive component.

7-6 Stereo Decoder Separation Measurement

The most basic test of a stereo decoder concerns the number of dB separation that it provides between left (L) and right (R) signals from a stereo signal generator. A typical generator is illustrated in Figure 7-15. Two classes of separation tests are made. First, an audio composite test signal may be applied to a decoder to determine the performance of the decoder alone. Second, an RF carrier may be frequency-modulated by the audio composite

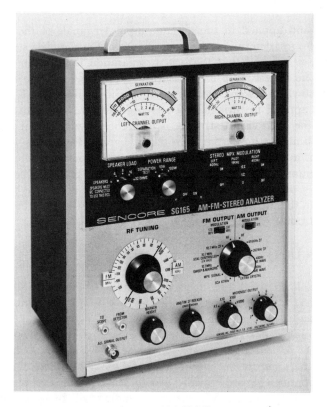

Figure 7-15. Appearance of an AM-FM-Stereo analyzer. (*Courtesy*, Sencore)

184 AUDIO MEASUREMENTS

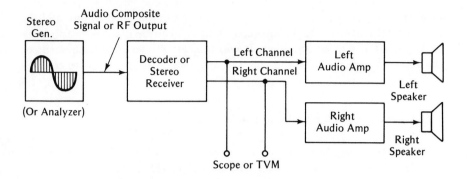

Note: FM-Stereo Analyzers Include dB Meters for Indication of Separation Values.

Figure 7-16. Test setup for measuring the separation between output channels of a stereo decoder.

test signal and applied to the antenna-input terminals of the FM stereo receiver. This test determines the performance of the complete stereo system. In general, it is found that the system separation is less than that of the decoder alone. A test setup for measuring the separation between L and R output channels of a stereo decoder is shown in Figure 7-16. L and R test signals are applied in turn from a stereo generator. When an L signal is applied in the separation test, there is theoretically zero output from the R channel and maximum output from the L channel. Conversely, when an R test signal is applied, there is theoretically zero output from the L channel and maximum output from the R channel. In practice, however, the operator will usually measure approximately 30 dB difference between the outputs from the two channels. A separaton of only 10 dB is interpreted as a borderline trouble symptom.

7-7 Audio Units

Some specialized units are utilized in audio technology. For example, volume units (VU) are utilized, in addition to dB and dBm units. A volume unit is a power ratio that indicates the level of a complex wave above a reference volume. A typical VU meter is illustrated in Figure 7-17. A VU measurement indicates the relative power level of a speech or music waveform. Volume units are not used to measure the power level of a sine-wave signal; dBm units are used for this purpose. If the power level of a sine-wave signal is measured with a VU meter, a reading will be obtained in dBm units. A VU measurement implies the application of a voice or music waveform, which is characterized by high peak values and a low average value. As a rule of thumb, it is commonly assumed that the average peak

7-7 AUDIO UNITS 185

Figure 7-17. A typical VU meter. (*Courtesy,* Simpson Electric Co.)

level in a program waveform is 10 dB above a sine-wave peak level. In practice, an audio system operating at a level of +12 VU will be tested for percentage of distortion at a sine-wave level of +22 dBm.

In VU measurements, the reference volume is specified as a strength of program wave that produces a reading of zero VU on a meter such as the one described above. This type of meter has specified damping, and is calibrated to indicate 0 VU with a sine-wave input that has a power level of 1 mW in 600 Ω. It follows that the term *reference volume* is not a precise concept and that it cannot be defined in fundamental terms. However, volume-unit measurements are of basic importance in the monitoring of audio systems in radio broadcast and television stations.

Next, the *phon* is a unit for measuring the apparent loudness level of a sound. It is numerically equal to the sound-pressure level, in decibels relative to 0.0002 microbar, of a 1-kHz tone that is considered by listeners to be equivalent in loudness to the sound under consideration. The relation of loudness units (phons) to frequency and to decibels is shown in Figure 7-18. Because the ear is not equally responsive to all frequencies, the phon unit and the dB unit are not identical. Note that dB values have no necessary relation to frequency. On the other hand, a phon value is a function of frequency. A frequency of 1 kHz has been stipulated as the common basis for dB and phon units, so that the level in dB is identical numerically with the level in phons at 1 kHz. It is assumed that the same zero reference level is employed in both situations.

At any frequency other than 1 kHz, the nonlinear response of the ear results in a difference between phon and dB values, as seen in Figure 7-18. A definition of the loudness level in phons may be stated in the form:

> The loudness of a sound in phons is numerically equal to the sound intensity in dB of an equally loud 1-kHz sine-wave tone.

In turn, it is possible to compare various sounds in loudness levels by

186 AUDIO MEASUREMENTS

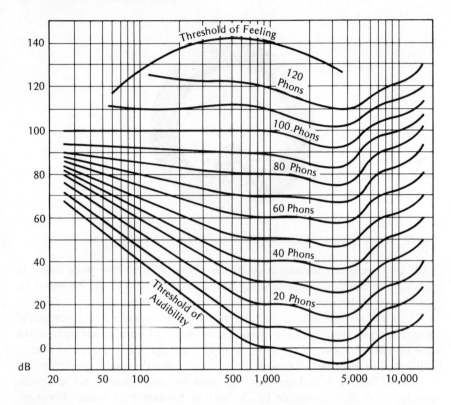

Figure 7-18. Relation of loudness units (phons) to frequency and to decibels.

comparing each with a 1-kHz sine-wave tone from an audio generator driving an earphone controlled by an attenuator calibrated in dB. The attenuator is adjusted so that the loudness from the phone at one ear is considered to be equally as loud as the sound entering the uncovered ear of the operator. The loudness level in phons is said to be numerically equal to the intensity of the reference tone in dB.

Next, consider the mel unit. This is a unit of pitch. The relation of mel units to hertz units is depicted in Figure 7-19. Thus, 1000 mels is stipulated as the pitch of a 1000-Hz tone; 500 mels as the pitch of a tone that sounds half as high; 2000 mels as the pitch of a tone that sound twice as high, and so on. It is evident from Figure 7-19 that perceptual evaluation of pitch is disproportional to the frequency of the tone that produces it. The foregoing relations apply to pure tones. On the other hand, when several frequencies are sounded together as a complex tone, the ear responds differently. As an illustration, if a complex tone consists of several frequencies that differ by a constant amount, the perceived pitch is frequently that of a tone with a

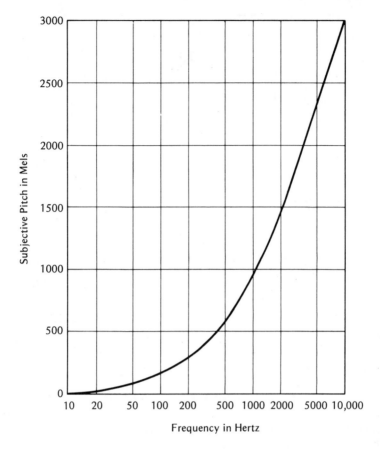

Figure 7-19. Relation of mel units to hertz units.

pitch equal to the common difference. For example, suppose that a complex tone has frequency components of 700, 800, 900, and 1000 Hz. The listener judges the pitch of the complex tone to be 100 Hz.

7-8 Tone-Burst Test of Speaker Enclosure

Tone-burst tests are generally used to check the performance of a speaker enclosure, as shown in Figure 7-20. A tone burst is a specialized type of waveform that is produced by 100 percent modulation of a sine wave by a square wave. The square-wave frequency is typically 100 Hz, with a sine-wave frequency of 1 kHz. To clearly indicate the presence of any spurious resonances, it is desirable to vary the sine-wave frequency over a wide range. Accordingly, a comprehensive test may employ sine-wave

188 AUDIO MEASUREMENTS

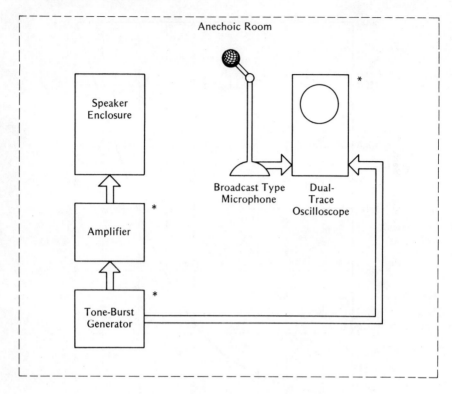

*Usually Installed Outside of Anechoic Room.

Figure 7-20. Test setup for checking speaker enclosure performance.

frequencies in the range from 100 Hz to 10 kHz. If a speaker system were ideal, the output from the broadcast-type microphone would be the same as the input waveform to the speaker. However, there is inevitably some residual distortion in the sound output from the enclosure, although it might be quite small. A dual-trace oscilloscope is helpful, so that a direct comparison of the input and output tone-burst waveforms can be made. Typical results of a tone-burst test on a high-quality speaker enclosure are illustrated in Figure 7-21. Observe that there are residual distortions present that produce slowed rise and a slight "hangover effect."

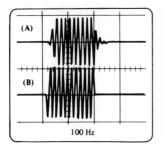
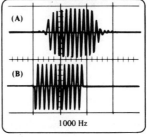
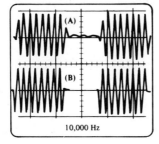

Figure 7-21. Example of a tone-burst test on a high-quality speaker enclosure. (*Courtesy,* Radio Shack, a Tandy Corp. Co.)

190 AUDIO MEASUREMENTS

Review Questions

1. Name several parameters that are measured in audio technology.
2. How is the frequency response of an audio amplifier measured?
3. Define the power bandwidth of an audio amplifier.
4. Discuss the measurement of harmonic distortion.
5. What is the distinction between intermodulation distortion and harmonic distortion?
6. Describe the characteristics of crossover distortion.
7. How does transient intermodulation distortion differ from conventional intermodulation distortion?
8. Explain how a music-power rating differs from an rms sine-wave power rating.
9. What parameters does an impedance bridge measure?
10. Why are capacitance standards utilized in inductance bridges?
11. Distinguish between a Maxwell bridge and a Hay bridge.
12. What is the standard test frequency employed by an impedance bridge?
13. How is stereo separation measured?
14. Describe the distinction between dB and VU units.
15. In what way does a phon unit correspond to a dB unit?

chapter eight

Digital Measurements

8-1 General Considerations

Digital measurements involve various kinds of test and measuring instruments, such as triggered-sweep oscilloscopes, pulse generators, logic probes and pulsers, digital frequency counters, logic clips and comparators, and conventional instruments such as digital multimeters. Oscilloscopes used in digital measuring procedures are comparatively sophisticated. Some of these instruments are combined *time-frequency domain* and *data-domain* indicators. Data-domain instruments, generally called logic state analyzers, are used to monitor binary digits (bits), digital words, digital addresses, and digital instructions as a sequence. The display is in binary form—1's and 0's arranged in columns and rows on the CRT screen. A time-frequency domain display of a digital clock signal, with typical distortions, is depicted in Figure 8-1. On the other hand, a data-domain display of digital events in a data stream is illustrated in Figure 8-2. This data-domain display is also called a *window* of digital events.

Electrical and functional analysis are not separable, but each is used to complement the other. As an illustration, only when word flow is incorrect as determined with a functional display need the operator be concerned with the voltage conditions that produced the digital words. Even when word-flow errors require electrical analysis, the number of signal nodes in the

192 DIGITAL MEASUREMENTS

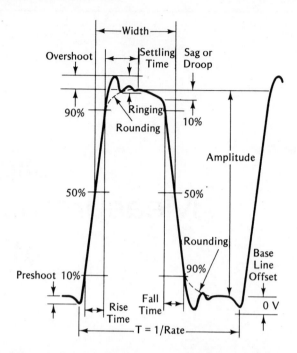

Figure 8-1. Time-frequency display of a digital clock signal, showing typical distortions. (*Courtesy*, Hewlett-Packard)

vicinity of the error complicates the use of oscilloscopes. Thus, it is helpful to define oscilloscope functions of probing, triggering, and display in terms of words versus event or sequence, or words versus time, rather than in volts versus time. The traditional analog picture of absolute versus sweep time provides careful analysis of electrical parameters. This is the case because the important information—amplitude versus time—is the information that the waveform carries. This method can help decipher noise (Figure 8-3), ringing, spikes, constant dc levels, voltage swings, and so on. Moreover, it is the analysis domain in which typical operators are most experienced and have the most confidence.

8-2 Digital Signal Characteristics

Digital information is often nonrepetitive. Extremely long (and fast) data sequences are common. Moreover, parameters that are significant for analog analysis are less important in a digital measurement. As an illustration, amplitude is usually important in that the voltage value must be either above or below threshold values (logic high or logic low). Furthermore, time is often unimportant in an absolute sense, although it becomes critical when related to the clock rate of a system. For example, a pulse may normally be

8-2 DIGITAL SIGNAL CHARACTERISTICS   193

delayed by a certain number of clock cycles. Figure 8-4 shows a typical delay of twenty clock cycles. Thus, a functional measurement consists of an observation of digital information (logic high or logic low) versus system time (clock). Accordingly a hierarchy of logic state test and measurement levels can be established. Each of these levels provides only the information necessary for that particular level of digital checkout. Thus, to effectively

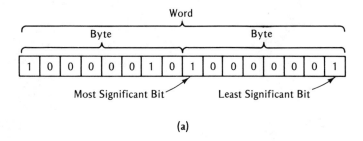

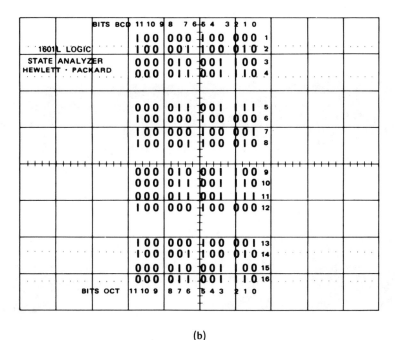

Figure 8-2. A logic-state analyzer is an oscilloscopic instrument that displays digital events in a data stream: (a) example of digital bytes and a digital word; (b) a data-domain display; (c) appearance of a logic state analyzer test setup. (*Courtesy, Hewlett-Packard*)

(c)

Figure 8-2. (Cont.)

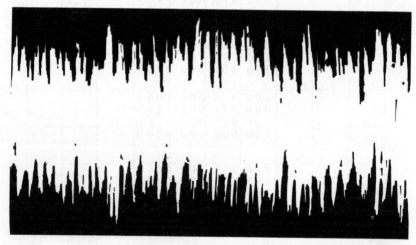

Figure 8-3. A typical noise waveform.

8-3 DIGITAL PROBE APPLICATION TECHNIQUES 195

cope with digital circuit operation, a logic state analyzer must meet several basic requirements:

1. Data must be read and presented in binary form for easy reading with no interpretation required.
2. A sufficient number of inputs should be provided so that the entire data word can be monitored at the same time.
3. A trigger point is required that is related to a unique data word within a sequence.
4. Digital delay is needed to position the display window to the desired point in time from the reference (trigger word).
5. Digital storage is needed to retain single-shot events along with negative time (data leading up to a desired trigger point).

Digital signals are almost invariably multiline and are difficult to interpret from a volts versus time display when the operator is concerned only with logic state versus system time. A typical logic state analyzer solves this problem by displaying digital words 32, 16, or 12 bits wide versus system clock in a table display that is very easy to use when the operator is examining functional relationships (refer back to Figure 8-2). The table displays are in terms of logic high's (1's) and logic low's (0's) versus a clock signal. Triggering is accomplished by using trigger-word switches that allow selection of a unique trigger point. Also, the display may be moved in system time from the trigger point by using digital delay in either a positive or a negative direction. An example of set-up time for a device is shown in Figure 8-5.

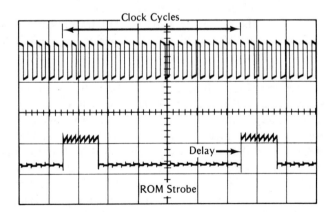

Figure 8-4. Dual-trace display illustrates a delay of twenty clock cycles between the upper and lower waveforms.

196 DIGITAL MEASUREMENTS

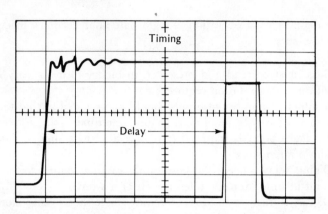

Figure 8-5. Example of setup time for a device. (*Courtesy, Hewlett-Packard*)

8-3 Digital Probe Application Techniques

The simplest form of data-domain test instruments consists of miniaturized probe-type pulse generators and indicators such as illustrated in Figure 8-6. A logic pulser is basically a single-shot pulse generator with a high output-current capability. It is generally used with a logic probe, or a logic current tracer, to provide an indication of pulse activity, faulty in-circuit IC's, and the static states of all pins. The circuit under test can be stepped one pulse at a time while the operator checks the truth tables of the logic packages in order to turn up any defects. In other words, a logic pulser provides a convenient means of injecting single pulses, the effects of which are monitored with a logic probe or with a logic current tracer. An example of a truth table for a 4-input AND/OR gate is shown in Figure 8-7. Note that the operation of a logic pulser is automatic and that no adjustments are required. The tip of the logic pulser is touched to an appropriate terminal in the circuit under test (Figure 8-6) and the pulse button is pressed. In turn, all circuits connected to the terminal (outputs as well as inputs) are briefly driven to their opposite state. No unsoldering of IC terminals is required.

An example of TTL logic levels is shown in Figure 8-8. The logic-high threshold is 2.4 V and the logic-low threshold is 0.4 V. If the voltage at the terminal under test falls within the interval from 0.4 to 2.4 V, it is defined as a "bad level", and the digital-logic probe will not respond. Note that the operator does not concern himself with whether the test point is logic-high or logic-low. High nodes are automatically pulsed low and low nodes are automatically pulsed high, each time the pulse button is pressed. The pulser illustrated in Figure 8-6 will source or sink up to 0.65 ampere, which is sufficient to override IC outputs in either the high or low state. An output pulse width of 0.3 μsec is employed to limit the amount of energy delivered

8-3 DIGITAL PROBE APPLICATION TECHNIQUES   197

Figure 8-6. Digital logic pulser and current tracer probes. (*Courtesy*, Hewlett-Packard)

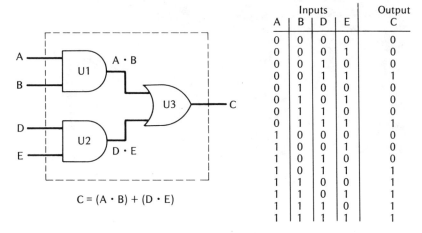

$C = (A \cdot B) + (D \cdot E)$

Inputs				Output
A	B	D	E	C
0	0	0	0	0
0	0	0	1	0
0	0	1	0	0
0	0	1	1	1
0	1	0	0	0
0	1	0	1	0
0	1	1	0	0
0	1	1	1	1
1	0	0	0	0
1	0	0	1	0
1	0	1	0	0
1	0	1	1	1
1	1	0	0	1
1	1	0	1	1
1	1	1	0	1
1	1	1	1	1

Figure 8-7. Arrangement of a 4-input AND/OR gate with truth table.

to the device under test, thereby eliminating the possibility of damage from excessive test energy. This type of logic pulser can be used with either TTL or DTL logic.

Consider the characteristics of the Hewlett-Packard multifamily logic probe illustrated in Figure 8-9. This type of probe simplifies and speeds up logic-circuit testing procedures. It indicates digital states and pulses in both

198 DIGITAL MEASUREMENTS

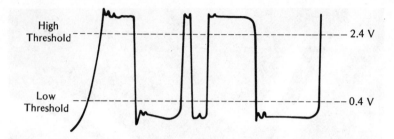

Figure 8-8. Voltage thresholds in a TTL signal. (*Courtesy, Hewlett-Packard*)

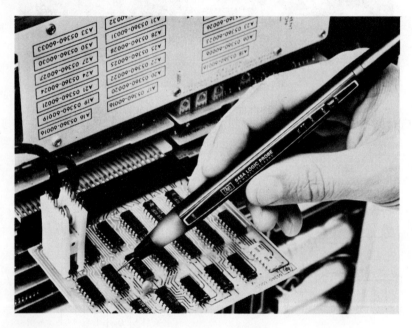

Figure 8-9. A multifamily logic probe. (*Courtesy, Hewlett-Packard*)

high-level (CMOS) and low-level (TTL) logic. A desired logic family is selected by means of a slide switch. The CMOS logic threshold levels are variable and are set automatically. This type of probe can be used with all positive logic up to +18 V, such as TTL, DTL, RTL, CMOS, HTL, HiNIL, NMOS, and MOS. Note that the current tracer illustrated in Figure 8-6 is an ultrasensitive digital logic tester that locates low-impedance faults by tracing the flow of current pulses rather than voltage changes in circuit conductors. By following the conducting path with the tip of this

8-3 DIGITAL PROBE APPLICATION TECHNIQUES   199

probe and watching the built-in indicator lamp, the operator can determine in-circuit logic current activity. With the ability to pinpoint one faulty point on a node, even on multi-layer boards, a current tracer can be used to locate faults such as solder bridges, short-circuited conductors in cables, short-circuits in voltage distribution networks, short-circuited IC inputs and outputs, stuck wired-OR circuits, and stuck data busses. It senses current pulses as small as 1 mA and as large as 5 mA from the conductor.

A logic probe is used to trace logic levels and pulses through integrated circuitry to determine whether the point under test is logic-high, logic-low, bad-level, open-circuited, or pulsing. A probe that is designed or set for logic levels of 2.0 and 0.8 V is utilized in TTL or DTL circuitry. When the probe is touched to a high-level point, a bright band of light appears around the probe tip. On the other hand, when the probe is touched to a low-level point, the light goes out. Open circuits or voltages that fall in the bad level region between the preset thresholds produce illumination at half brilliance. Single pulses of 10 ns or greater widths are made readily visible by stretching them up to one-twentieth of a second. The probe light flashes on or blinks off depending upon the polarity of the pulse. Pulse trains up to 50-MHz repetition rate cause the lamp to blink off and on at a 10-Hz rate.

The circuit under test can first be operated at normal speed while checking for the presence of key signals, such as clock, reset, start, shift, and transfer pulses. Next, the circuit can be stepped one pulse at a time while the operator checks the truth tables of the logic packages in order to turn up any defects. False triggering is sometimes caused by noise glitches (malfunctions) as illustrated in Figure 8-10. When it is suspected that a glitch is present, a follow-up test is required with a high-speed oscilloscope, preferably of the storage type. Narrow glitches are often difficult to display on other types of oscilloscopes, even when the operator is certain that they are present. Again,

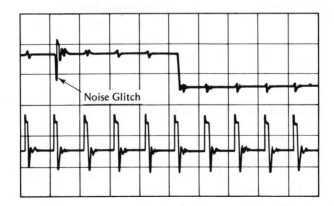

Figure 8-10. A noise glitch caused by a power-supply surge. (*Courtesy*, Hewlett-Packard)

200 DIGITAL MEASUREMENTS

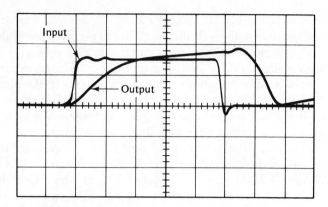

Figure 8-11. Example of a substantially distorted output pulse.
(*Courtesy*, Hewlett-Packard)

an output pulse sometimes becomes distorted (see Figure 8-11). This distortion involves excessively slow rise time and delay. One typical result of slow rise time is production of a "race" condition. This is a type of malfunction in which a signal propagates through two or more memory elements during one clock period. As before, a follow-up test is required with a high-speed oscilloscope with calibrated time bases whereby the rise time of the suspected pulse is measured to determine if it is within normal limits.

High-level logic such as HTL, HiNIL, MOS, discrete, and relay requires probe thresholds of 2.5 and 9.5 V. HTL denotes high-threshold logic; HiNIL denotes high noise-immunity logic. Discrete logic employs individual components and devices. Relay logic is electromechanical in design and does not employ semiconductor devices. Emitter-coupled logic (ECL) requires logic thresholds of -1.1 and -1.5 V. An ECL logic probe contains high-speed circuitry that stretches single-shot pulses as narrow as 5 ns sufficiently to be clearly displayed. As noted previously, all logic probes find chief application in preliminary analysis procedures. After preliminary evaluation of circuit response has been completed, it is often desirable or necessary to follow up with oscilloscope tests.

8-4 Oscilloscope Waveform Analysis

Operation of an ECL probe is correlated with waveform analysis as depicted in Figure 8-12. If the signal level at the probe tip is greater than -1.60 V, the indicator light at the end of the probe is dark. Next, if the signal rises to less than a -1.0 V level, the indicator light glows. Even a very brief fall of the signal voltage to more than a -1.60 V level will cause the light to blink off, and then remain on. Conversely, if the signal level rises very briefly from a high level to less than -1.0 V, the light will blink on and then remain off.

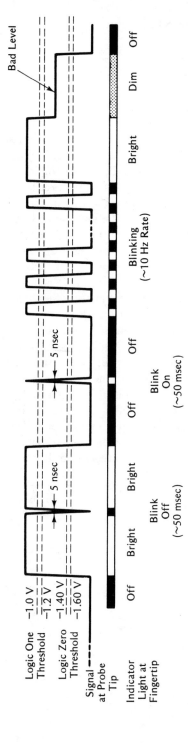

Figure 8-12. Response of an ECL probe to various input signals. (*Courtesy*, Hewlett-Packard)

202 DIGITAL MEASUREMENTS

When a pulse train is applied to the probe tip, the indicator light blinks on and off at a 10-Hz rate. In the event that the probe is applied at a bad-level point in the circuit under test, the indicator light glows at half brilliance. Accordingly, the logic probe provides some of the basic information that is usually considered to be the province of the oscilloscope.

It is helpful to define oscilloscope functions of probing, triggering, and display in terms of words versus event or sequence, or words versus time, rather than in terms of volts versus time. The traditional analog picture of absolute voltage versus sweep time provides precise analysis of electrical parameters. This is the case because the essential data—amplitude versus time—is the information carried by the waveform. It assists the operator in evaluation of noise, ringing, spikes, constant dc levels, voltage swings, and so on. An idealized correlation of digital waveforms with LED indication on a logic state analyzer is shown in Figure 8-13. Observe that each LED indication is synchronized with a clock pulse. Whenever the digital signal is high, the corresponding LED glows; on the other hand, when the digital signal is low, the corresponding LED is dark.

Observe the timing levels along the leading and trailing edges of the trigger pulse for a JK flip-flop as shown in Figure 8-14. First, as the leading edge crosses the logic-low level, the flip-flop responds to isolate the slave section from the master section. Second, as the leading edge crosses the logic-high level, the flip-flop responds by enabling the J and K inputs to set the master section. Third, as the trailing edge crosses the logic-high level, the flip-flop responds by disabling the J and K inputs. Fourth, as the trailing edges crosses the logic-low level, the flip-flop responds by transferring the data from the master section to the slave section. It is evident that if a trigger

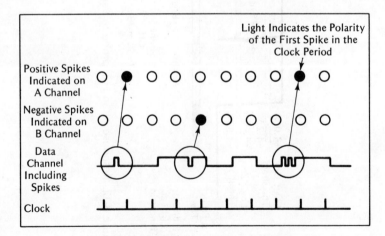

Figure 8-13. Correlation of digital waveforms with LED indication on a logic state analyzer. (*Courtesy*, Hewlett-Packard)

8-4 OSCILLOSCOPE WAVEFORM ANALYSIS   203

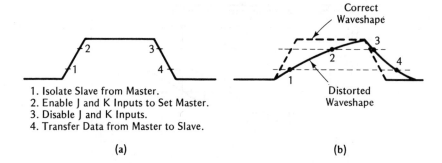

Figure 8-14. Timing levels on the trigger pulse of a JK flip-flop: (a) ideal waveshape; (b) distorted waveshape.

pulse is distorted, for example, so that the second step is delayed and occurs simultaneously with the third step, the JK flip-flop will malfunction accordingly.

Elaborate oscilloscopes designed for digital equipment analysis often combine LED readout at the bottom of the CRT screen, as illustrated in Figure 8-15. In this example, the readout indicates the time from one leading edge of a pulse to that of the next pulse, from t_1 to t_2. The elapsed time is 1.92 μsec. A microprocessor section is included in the oscilloscope to measure the period between the markers and to drive the LED readout. For increased measurement accuracy, the oscilloscope is operated in its delayed-sweep mode in order to display the two intensified traces alternately, as illustrated in Figure 8-16. The controls are then adjusted to make the two

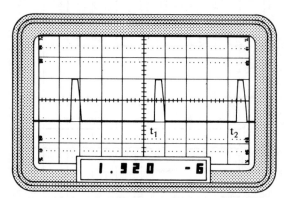

Figure 8-15. Two intensified markers are operator-positioned to cover the start and stop points of the interval to be timed. Then, the LED readout automatically and continuously displays the time between the two markers (192 μsec). (*Courtesy*, Hewlett-Packard)

204 DIGITAL MEASUREMENTS

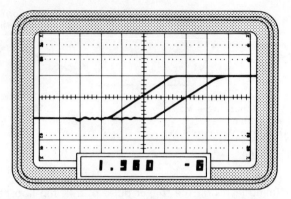

Figure 8-16. Maximum readout accuracy is obtained by superimposing the two traces and adjusting their leading edges to coincide. (*Courtesy*, Hewlett-Packard)

traces coincide, whereupon the readout develops maximum accuracy of indication (in this example, 1.962 μsec). Greater accuracy results from the circumstance that the operator can better judge the coincidence of the leading edges of the waveforms than marker placements can be positioned.

Consider next the display of a race condition pulse (glitch) as shown in Figure 8-17. In many digital circuit applications, it is necessary to use external trigger sources to maintain proper timing relationships and to know

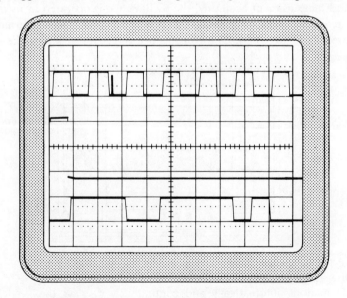

Figure 8-17. An analog display of digital data that shows a race condition pulse (top trace) which is defined in time by the third channel trigger view. (*Courtesy*, Hewlett-Packard)

8-4 OSCILLOSCOPE WAVEFORM ANALYSIS 205

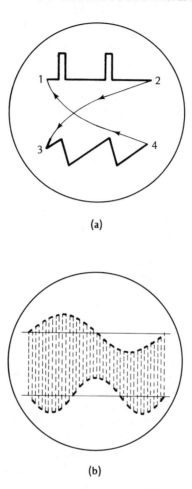

Figure 8-18. Basic dual-trace display modes: (a) alternate-channel mode; (b) chopped mode. (*Courtesy,* Sencore)

the time relationship of the trigger signal to the displayed events. An elaborate oscilloscope may be designed with a "trigger view" pushbutton. If this pushbutton is depressed while the oscilloscope is operating in either its alternate or chopped mode, the external trigger signal is displayed as a third channel with the trigger threshold at center screen. By adjusting the trigger-level control on the oscilloscope, the operator can see which portion of the trigger signal is initiating the sweep.

The distinction between alternate-channel and chopped mode displays is seen in Figure 8-18. In other words, the first waveform is completely traced on the screen, and then the second waveform is completely traced on the screen below the first waveform in the alternate-channel mode of dual-trace display. On the other hand, both the first and the second traces are traced by discontinuous intervals, with the beam alternating rapidly between the two traces in the chopped mode of dual-trace display. Both modes have advantages and disadvantages, and most dual-trace oscilloscopes provide a choice of mode. The alternate-channel mode finds most satisfactory utility in high-speed sweep patterns, whereas the chopped mode is preferred in low-speed sweep patterns. When high-speed chopping action is employed in low-speed sweep patterns, the traces appear to be almost continuous.

8-5 Digital Delay Techniques

It was previously noted that a digital display (Figure 8-2) can be moved on the screen of a logic state analyzer from the trigger point by using digital delay in either a positive or a negative direction. Negative digital delay is provided by the inherent storage features of logic analyzers. This feature permits the instrument to display a number of events leading up to a selected trigger event. A typical logic analyzer can display up to 64 bits of serial-A mode data that occur before the trigger point. Positive delay allows movement of the display downstream from the trigger. As a case in point consider a disc memory. The start of a sector may be its only available unique trigger point, yet, the data to be analyzed may be located thousands of bits downstream from the trigger. An analyzer with digital display can position the display window precisely at the exact location of the character or signal to be examined. In digital systems, very low repetition rate or single-shot events are encountered that require storage to permit analysis. As an example, "once-per-keystroke" calculator sequences fall into this category. Logic state analyzers contain sufficient memory to capture and store such events and are in turn highly useful in single-shot operation analysis.

Digital triggering and delay are necessary for functional analysis. They are also of great value when the operator is "aiming" or positioning electrical analysis windows on oscilloscopes. These capabilities are needed for both serial and parallel data stream analysis. They allow the operator to "window in" on events that occur as part of very long data sequences. In serial data analysis, the problem of data pattern recognition can be solved if the data or instruction portions of a serial word are known. It then becomes possible to generate a unique trigger word from a known serial event. As an illustration, if a pattern set on a logic state analyzer matches the bits contained in the instruction portion of a serial word, a trigger is generated. Thus, a unique trigger is defined to allow analysis of serial data streams. Added to this is the capability of digital delay, which allows further indexing from the operator-selected trigger point. (Refer to Figure 8-19.)

For parallel data analysis, it is often necessary to trigger on the simultaneous occurrence of several events. As an illustration, if one or more

8-5 DIGITAL DELAY TECHNIQUES   207

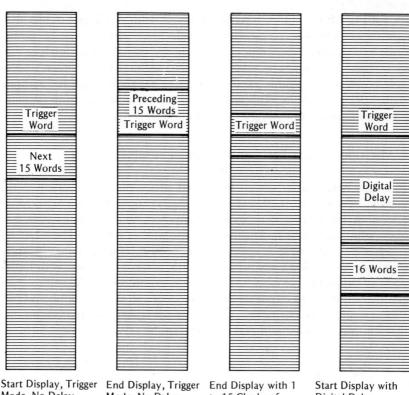

Start Display, Trigger Mode, No Delay. In this Mode the Trigger Word and the Next 15 Words are Displayed.

End Display, Trigger Mode, No Delay. The 15 Words Preceding the Trigger Word are Displayed.

End Display with 1 to 15 Clocks of Digital Delay. By Selecting End Display and Digital Delay up to 15, it is Possible to Simultaneously Display Words Occurring Before and After the Trigger Word.

Start Display with Digital Delay. The Trigger Word is Not Displayed in this Mode. After the Preselected Delay a Display Field of 16 Words is Displayed.

Figure 8-19. Examples of digital delay features. (*Courtesy, Hewlett-Packard*)

channels of data go high at the same point in time that the clock signal goes high, a trigger could be generated at this point. Moreover, the selected trigger events could be either high- or low-polarity signals. Triggering need not be clock-related; it may be asynchronous. This feature allows the operator to initiate the display sequence on a signal that might not be present when the clock samples the inputs to the analyzer. Signals such as spikes, or other random events, can accordingly be detected or used as trigger events. Logic state analyzers are supplemented by trigger probes that feature TTL, MOS, and ECL compatibility. A 4-bit AND gate trigger with selectable bit levels is commonly provided. These circuit-powered probes provide 4-bit pattern recognition triggering for digital signal analysis and may be used for both functional and electrical analysis. Another type of trigger probe provides 8-bit parallel triggering capability with the addition of digital delay capability. This facility provides versatile triggering capabilities for oscilloscope windowing to digital problem areas.

8-6 Microprocessor Tests and Measurements

A microprocessor is a large-scale integrated circuit that is equivalent to the central processing unit (CPU) of a large digital computer. Microprocessors are used in large computers, in minicomputers, and in microcomputers. Large computers can be designed as multiprocessing systems. These systems are organized with two or more interconnected computers that perform functionally specialized tasks. In other words, a large computer with multiprocessing capability contains two or more microprocessors. A fault-tolerant digital system employs duplicate microprocessors, so that if one should fail, the other will be automatically switched into the system. Microprocessors are utilized in minicomputers to perform separate functions that were formerly

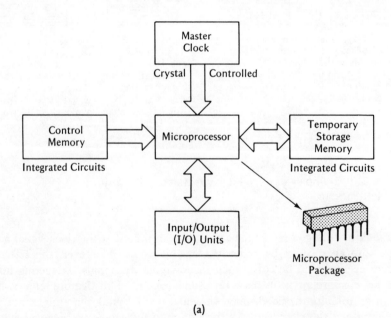

(a)

Figure 8-20. Minimal general-purpose computer functional diagram: (a) system block diagram; (b) microprocessor organization.

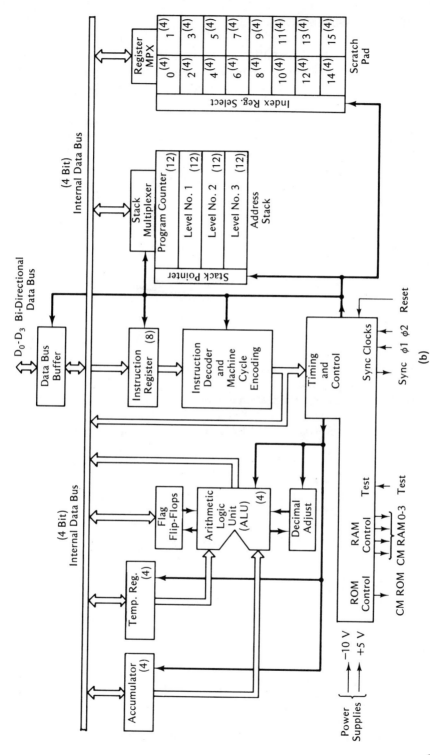

Figure 8-20. (Cont.)

assigned to a single elaborate CPU. A microcomputer comprises a microprocessor, a control memory, a temporary storage memory, and a master clock (see Figure 8-20). This arrangement is termed a minimal general-purpose computer.

A widely used microprocessor, and the first commercial device, is the Intel 4004 microprocessor. It operates from $+5$ and -10 V power supplies and features a 4-bit parallel CPU with 46 instructions. This microprocessor can directly address 4K 8-bit instruction words of program memory and 5120 bits of data storage random-access memory (RAM). Up to sixteen 4-bit input ports and sixteen 4-bit output ports can also be directly addressed. Sixteen index registers are provided internal to the microprocessor for temporary data storage. This microprocessor operates at clock rates up to approximately 750 kHz. Pin assignments for the 4004 microprocessor are shown in Figure 8-21.

Data and control lines for the microprocessor are summarized as follows:

D0–D3; bidirectional data bus for handling all address and data communication between the microprocessor and the RAM and ROM (Read Only Memory) chips.

$\phi 1$–$\phi 2$: nonoverlapping clock signals that determine the timing of the microprocessor.

Sync: synchronization signal that indicates the beginning of the instruction cycle to the RAM and ROM chips.

Reset: a "1" level applied to RESET clears all flag and status flip-

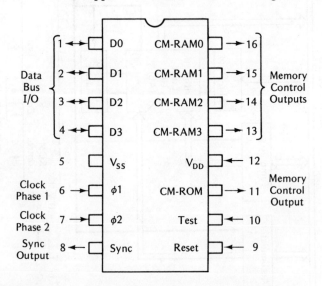

Figure 8-21. Pin assignments for the Intel 4004 microprocessor.

flops and forces the program counter to 0. RESET must be applied for 64 clock cycles (8 machine cycles) to completely clear all address and index registers.

Test input: the logic state of TEST can be examined with JCN (Jump Conditional) instruction.

CM-ROM: line enables a ROM bank and I/O devices that are connected to the CM-ROM line.

CM-RAM0 through CM-RAM3: lines function as bank select signals for the RAM chips in the system.

Test Probe Connections

Tests of the microprocessor system are greatly facilitated by real-time analysis of program flow, triggering on specific events, and so on. The Hewlett-Packard data domain analyzers such as illustrated in Figure 8-2 are particularly well adapted for this purpose. The 4004 microprocessor does not provide a unique clock for the logic state analyzer at the proper time (end of A3 state) in the instruction cycle. The CM-ROM line is always true at A3

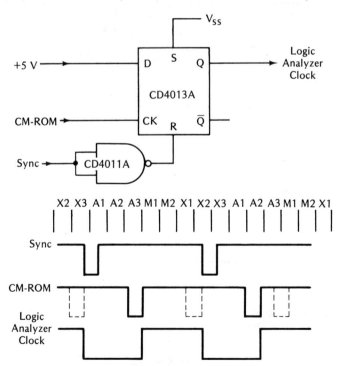

Figure 8-22. Circuit for deriving a clock for the HP 1600A from the 4004 sync and CM-ROM signals. (*Courtesy*, Hewlett-Packard)

212 DIGITAL MEASUREMENTS

and it can be used as a clock signal. However, CM-ROM also occurs at states M2 or X2 during the execution of some instructions, which would result in invalid data being displayed by the analyzer. A correct display can be ensured by constructing the circuit depicted in Figure 8-22.

If the portion of the program to be examined is completely contained on one ROM chip, the chip-select line (CS) for that ROM can be used as a clock. The probe connections shown in Figure 8-23 provide a display of the activity on the address line. A system that will not "come up" can frequently be debugged by monitoring address flow alone. The 4004 CPU chip has a

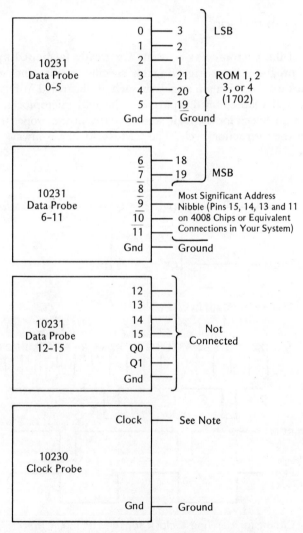

Figure 8-23. Probe connections. (*Courtesy*, Hewlett-Packard)

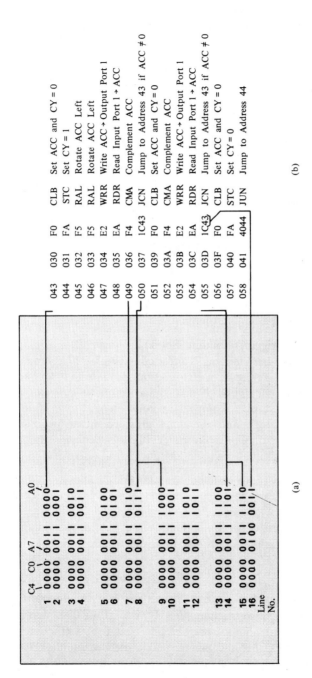

Figure 8-24. System response to test routine: (a) real-time state analysis; (b) cross-assembler listing output.

214 DIGITAL MEASUREMENTS

4-bit data bus, on which the 12-bit address is multiplexed during A1, A2, and A3 states of the 4004 machine cycle. To view the demultiplexed 12-bit address on the 1600A, the 4004 system must use 4008/4009 Standard Memory and I/O Interface Set, the 4289 Standard Memory Interface, or similar logic circuits that provide a demultiplexed address bus. If the system employs memory chips that decode the multiplexed address internally, such as the 4001 ROM, the microprocessor data should be monitored as detailed subsequently.

To set the controls, the operator turns the power on and sets the logic state analyzer controls as follows: Display Mode, Table A; Sample Mode, SGL. Note that SGL is selected for viewing single-shot events. The operator presses RESET to start the system. The first time that the system passes through the trigger point, the display will be generated and stored. For programs that are looping or cycling through the selected address, select REPET sample mode. The Start Display is set to ON; Trigger Mode is set with NORM/ARM to NORM, LOCAL/BUS to LOCAL, and OFF/WORD to WORD. The threshold is set to VAR, and is adjusted to 3.7 V. In the case of TTL-compatible systems, the threshold is set to TTL. The LOGIC control is set to POS; CLOCK, to $\sqcap$, and all other pushbuttons are set to their Out Position. The Display Time control is set ccw; Qualifiers, OFF; Trigger Word switches are set to the address at which the operator wishes to trigger; Column Blanking is set after the display is on-screen, whereupon the blanking is adjusted to display 12 columns of data.

Next, consider the display interpretation. Assume that a segment of a chip tester program for Quad NAND gates is to be examined. Proper operation is confirmed by a comparison between real-time state analysis shown in Figure 8-24(a), and the 4004 cross-assembler program listing output depected in (b). The chip tester routine performs the following functions:

1. Sets up bit patterns in the accumulator.
2. Outputs the accumulator contents to the NAND gates that are connected to I/O port 1.
3. Reads the gate outputs.
4. Tests on the gate outputs and indicates whether the chip is good or bad.

Consider the program listing in Figure 8-24(b). The instructions located in addresses 030 through 033 load the bit pattern 0010 into the accumulator. The next instruction, WRR in location 034, writes the accumulator contents into output port 1. The next two instructions (address locations 035 and 036) read the gate outputs present at input port 1 into the accumulator and complement the accumulator. Examination of lines 1 through 7 of the state display photograph in Figure 8-24(a) shows that these instructions have been executed in the proper sequence.

8-6 MICROPROCESSOR TESTS AND MEASUREMENTS   215

The instruction starting at address 037 is a conditional jump which is a 2-word instruction (lines 8 and 9 of the state display). If the chip passed the test—accumulator contains all zeros—the program continues the test routine. If the chip failed the test, the program jumps to an output routine. Examination of line 10 of the state display, address 039, reveals that the chip passed the test. The program then outputs another bit pattern (1111) to the chip under test and reads the input port. This is shown by lines 10 through 13 of the state display. Lines 14 and 15 of the state display are the addresses of the two words of another JCN instruction. Line 16 of the state display is the address 043, showing that the chip failed the last test, causing the program to jump to the output routine.

Consider next the map display. If a tabular display is not presented in the foregoing procedure, it means that the system did not access the selected address, and the No Trigger light will be on. To find where the system is residing in the program, the operator switches to "map" as illustrated in Figure 8-25. Using the Trigger Word switches, he (or she) moves the cursor (an illuminated circle) to enclose one of the dots on the screen. Then he switches to Expand and finalizes the cursor position. The No Trigger light will then go out and switching back to Table A will display the 16 addresses around that point.

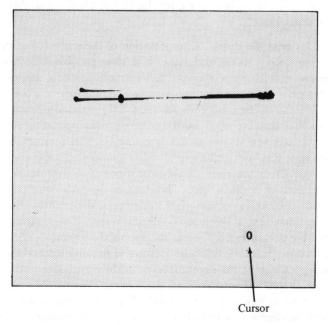

Figure 8-25. A map display shows the entire system activity. (*Courtesy,* Hewlett-Packard)

216 DIGITAL MEASUREMENTS

When program deviations occur, the reason may be as simple as a program error, or as complicated as a hardware failure on the data bus or command lines. Additional input channels now become very desirable. By combining the instruments illustrated in Figure 8-2, the trigger and display capability can be expanded to 32 bits wide, allowing the 12-bit address, 8-bit data word, and up to 12 other active control signals to be viewed simultaneously. The operator procedure is as follows:

1. Connect data cable between rear panel connectors.
2. Connect trigger bus cable between front panel bus connectors.
3. Set the 1600A controls as explained previously, with the following exception: set the display mode to TABLE A & B.
4. The 1607A controls are set as follows: Sample Mode, SINGLE; Start Display, ON; Trigger Mode with NORM/ARM to NORM; LOCAL/BUS to BUS; OFF/WORD to OFF; Threshold, Logic, and Clock as noted previously; all other pushbuttons to their Out positions; Qualifiers Q, Q0 to OFF.
5. Connect the data and clock inputs for the 1607A as follows: Connect 1607A data inputs 0 through 7 to the demultiplexed data bus (ROM output) starting with LSB connected to 1607A data input 0; Connect 1607A clock input to signal used to clock the 1600A; Connect grounds to appropriate points.
6. After a display is on-screen, set the 1607A blanking to display eight columns.

Consider next the display interpretation of the address and data lines. By displaying both address and data, it is now possible to confirm exact system operation with respect to the test routine. Looking at line 1 of the state display photograph in Figure 8-26, observe that the data corresponding to address 030 is F0, the 8-bit code for the CLB instruction. Looking at line 2, it is seen that the displayed word agrees with the operation code for the STC instruction given in the program listing. In this manner, subsequent lines of the state display can be examined to show exact program operation. Note that line 14 of the state display corresponds to the first word of the JCN instruction at address 03D. The data in line 15 corresponds to the second word (0100 0011) of the JCN instruction, the address that program control is transferred to if the jump condition is true. Examination of line 16 reveals that the program did jump to the specified address.

It is instructive to consider the viewing of the multiplexed data bus. In the preceding examples, the demultiplexed address and data lines have been observed on the ROM address and data lines. When a hardware failure occurs, however, it may be very helpful to directly observe activity on the multiplexed microprocessor data line. In the following example, observe that the data are being demultiplexed into a 12-bit address for driving a ROM.

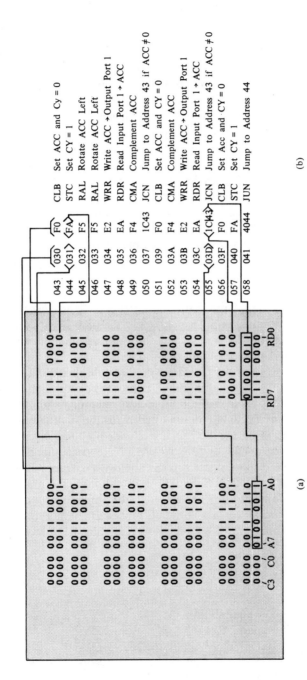

Figure 8-26. System response to test routine on address and data lines. (*Courtesy*, Hewlett-Packard)

The operator watches the ROM output being multiplexed back onto the data bus. The 1600A and 1607A are set up as follows to obtain the display:

1. Set the 1600A data, qualifier, and clock input with 1600A data inputs 0 through 7 to RD0 through RD7 on the ROM in order; set 1600A data inputs 8 through 15 to A0 through A7 in order; set 1600A Q0 input to ROM 0 chip select line (1702A, pin 14).
2. Note that by qualifying on $\overline{CS}$ and triggering on A0 through A7, the operator derives a unique trigger that is effectively 12 bits wide, with only the eight least-significant bits displayed.

The 1607A data and clock inputs are connected to the 4004 microprocessor as follows: (a) 1607A data inputs 0 through 3 to D0 through D3; (b) 1607A data input 4 to CM-ROM; (c) 1607A data input 5 to SYNC; (d) 1607A clock input to $\phi 2$. Set the 1600A controls the same as explained previously, with the following exceptions: Display Mode, Table A + B; End Display, ON; Delay, ON with Delay set to 8; Qualifier, TRIG with Q0 set to LO; Column blanking, ccw.

Set the 1607A controls as follows: End Display, ON; Delay, ON with Delay set to 8; Logic, Neg. Note that the microprocessor uses negative logic; i.e., the most positive voltage is a logic "0", and the most negative voltage is a logic "1". After a display is obtained, the operator adjusts the 1607A column blanking to display 6 columns in Table B.

Next, consider display interpretation of the multiplexed data bus. The state display in Figure 8-27 shows a comparison of the demultiplexed address and data buses (Table A) with the multiplexed microprocessor bus (Table B). Compare line 8 of Table A (trigger word) with the multiplexed data in Table B. Examination of the SYNC line shows that line 6 of Table B corresponds with instruction cycle state A1. Note that the SYNC and CM-ROM pulses are displayed as 1's in the photograph, since negative logic has been selected on the 1607A. Comparison of states A1, A2, and A3 (lines 6, 7, and 8 of the Table B state display) with the trigger word address bits reveal that the demultiplexer has correctly processed the address from the 4004. Similarly, comparison of trigger word data bits RD7 through RD0 with states M1 and M2 (lines 9 and 10 of the ROM display) shows that the multiplexer has correctly processed the ROM data onto the 4004 data bus. Observe that the CM-ROM line is true during the M2 state, indicating that the instruction being executed is an I/O instruction.

From the foregoing examples, it may be concluded that efficient debugging of the 4004 microprocessor system is expedited by two factors: First, the availability of the program listing as produced by the 4004 cross-assembler, and second, the availability of real-time logic state analysis for rapid error detection and correction.

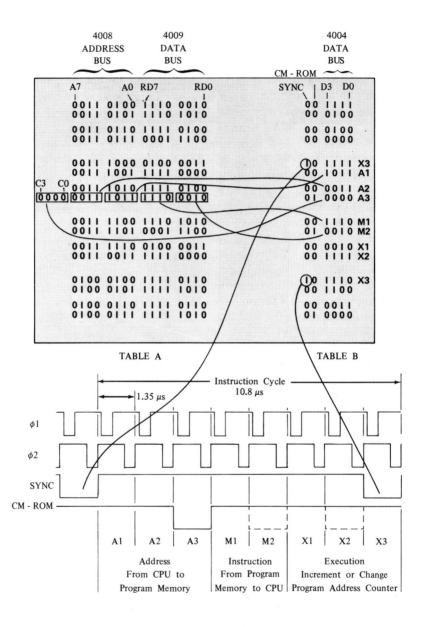

Figure 8-27. Comparison of 4004 data bus activity with demultiplexed address and data. (*Courtesy*, Hewlett-Packard)

Testing of the 8-Bit Microprocessor System

The Intel 8008 8-bit microprocessor chip has more computing capability and flexibility than the 4004. It is more suitable for control applications and data handling. The 8008 microprocessor family operates from +5 and −9 V sources. This microprocessor features an 8-bit address and data bus (D0 through D7) that, by time multiplexing, allows control information, 14-bit addresses, and 8-bit data bytes to be transmitted between the CPU and the external memory. Pin assignments are depicted in Figure 8-28. The 14-bit address permits direct addressing of 16K words of memory. This microprocessor provides state signals, cycle control signals, and a synchronizing signal to peripheral circuits. These lines are decoded externally to the microprocessor to provide the control and timing signals for the microprocessor system. All microprocessor inputs are TTL compatible and all outputs are low-power TTL compatible. The microprocessor operates with a 500-kHz clock.

Control lines for the 8008 microprocessor are summarized as follows: When the interrupt (INT) lines are enabled (HIGH), the CPU recognizes the interrupt request at the next instruction fetch cycle. RDY: HIGH (Logic "1") indicates to the CPU that valid memory data are now available. LOW

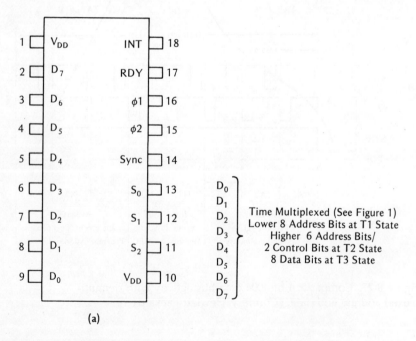

(a)

Figure 8-28. Intel 8008 microprocessor: (a) pin assignments; (b) microprocessor organization.

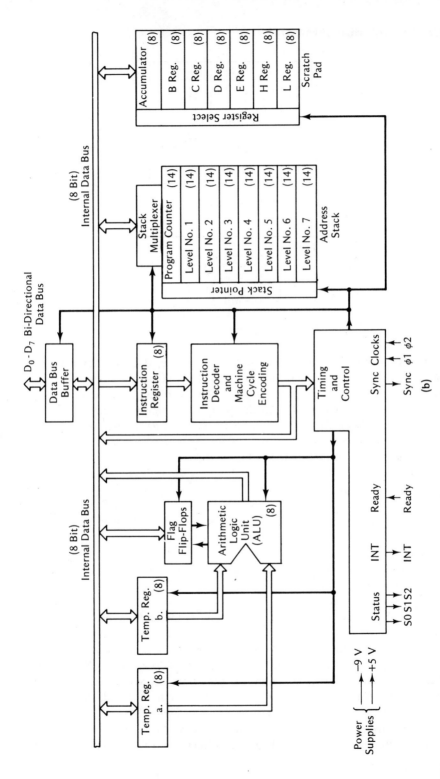

Figure 8-28. (Cont.)

S_0	S_1	S_2	State
0	1	0	T1
0	1	1	T11
0	0	1	T2
0	0	0	Wait
1	0	0	T3
1	1	0	Stopped
1	1	1	T4
1	0	1	T4

Figure 8-29. Tabulation of state control signals.

D_6	D_7	Cycle
0	0	Instruction Fetch Cyle (PCI)
0	1	Data Read (PCR)
1	0	I/O Operation (PCC)
1	1	Data Write (PCW)

Figure 8-30. Tabulation of cycle control bits.

(Logic "0") indicates to the CPU that valid memory data are not available. SYNC: Synchronizing signal indicating the start of each machine state. S0, S1, S2: State control signals. S0, S1, and S2 control use of the data bus and indicate the state of the CPU to the peripheral circuitry. These states are tabulated in Figure 8-29. D6, D7: Cycle control bits, designating whether cycle is instruction fetch, data read, data write, or I/O operation, at T2 state. These cycles are tabulated in Figure 8-30.

Test Probe Connections

A system that will not "come up" can often be debugged by monitoring the address flow alone. Since the 8008 14-bit address is time multiplexed, external address latches, such as the Intel 3404 latch, are required in an 8008 system. Connect the analyzer probes to the output side of the eight LSB address latches and to the input side of the MSB address latches and the cycle control bit latches. The analyzer probe connections shown in Figure 8-31 provide a display of the activity on the address lines. To set the analyzer controls, the operator turns the power on and proceeds as follows: Display Mode, Table A; Sample Mode, REPET. Note that if a program is not looping or cycling through the selected address, select SGL, press RESET, and start the system. The first time that the system passes through the trigger point, the display will be generated and stored. The timing diagram for the 8008 microprocessor is shown in Figure 8-32.

The trigger Mode is set with NORM/ARM to NORM, with LOCAL/BUS to LOCAL, and OFF/WORD to WORD. START DISPL

8-6 MICROPROCESSOR TESTS AND MEASUREMENTS 223

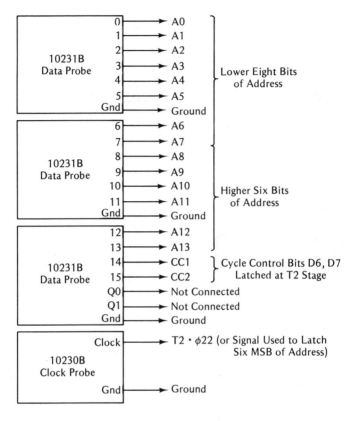

Figure 8-31. Data and clock probe connections. (*Courtesy,* Hewlett-Packard)

is set to ON; CLOCK is set to ⎍. In the system used for this example, the higher 6-bit address latches and cycle code bit latches are clocked on the leading edge of T2 · ϕ22. Note that the clocking requirements of another 8008 system may vary from this example. THLD is set to TTL; the 8008 output lines are low-power TTL compatible, as are most address latches. If other logic levels are used, set THLD to VARIABLE and adjust the threshold to match the given threshold level. All other pushbuttons are set to their Out positions; Display Time, ccw; Column Blanker, ccw; Qualifier Q1, Q0 is set to its OFF position; Trigger Word switches are set to match the address on which the operator wishes to trigger.

Consider next the interpretation of the display. In this illustration, system response to a CALL instruction is explained. The CALL instruction starts a subroutine to check the keyboard for the presence of a STOP command and to check the system status. Proper operation is confirmed by a comparison between real state time analysis, Figure 8-33(a), and the 8008

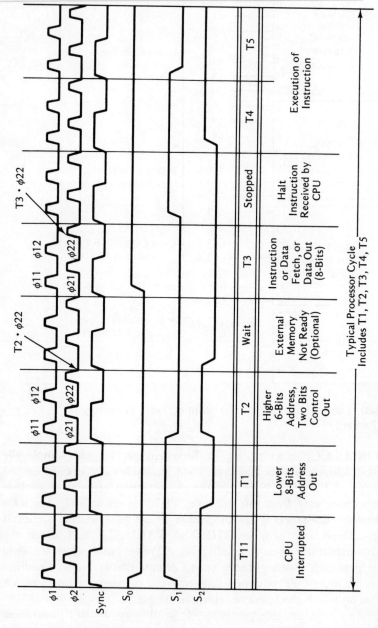

Figure 8-32. Timing diagram for the 8008 microprocessor. (*Courtesy*, Hewlett-Packard)

8-6 MICROPROCESSOR TESTS AND MEASUREMENTS 225

cross-assembler listing output, in (b). The 8008 responds to a CALL instruction in the following manner:

1. Store the content of the program in the push-down address stack.
2. Jump unconditionally to the instruction located in the memory site addressed by byte two and byte three of the CALL instruction.
3. Begin execution of subroutine.

Note the program listing depicted in Figure 8-33(b). Observe the 3-byte CALL instruction at location 00400. The first byte is the operation code, indicated by 00 in bits 15, 14 columns. The second and third bytes form a double-byte operand (indicated by 01 in bits 15, 14 columns), in this case the address of the first instruction in the subroutine. Proper operation of the CALL instruction is confirmed by observing that the address immediately following the third byte of the CALL instruction, 00402, is 00572. This means that the microprocessor fetched 172 (lower 8 bits of subroutine address) from location 00401 and 01 (higher 6 bits of subroutine address) from location 00402.

$$00\ 000\ 101\ 111\ 010 \leftarrow (000\ 001), (01\ 111\ 010)$$

The MVI H, KYBRD, and MVI L, KYBRD instructions (Load Keyboard address in H and L registers) may be confirmed by observing the fourth, fifth, sixth, and seventh lines of the table display photograph. Line 4 is the fetch of the MVI H operation code and line 5 is the fetch of the higher six bits of the keyboard address. Line 6 is the fetch of the MVI L operation code with line 7 being the fetch of the lower eight bits of the keyboard address. Line 8 is the fetch of operation code for MOV A, M and line 9 is the fetch of the keyboard character. In a similar manner, each instruction in the subroutine may be shown to have been properly executed.

To view addresses following the last displayed address, simply set the Trigger Word switches to match the address displayed in line 16. This address becomes the trigger word in line 1 with the next 15 addresses listed in lines 2 through 16. If the operator wishes to retain the original trigger point, an alternate technique is to use digital display and to set the thumbwheels to 00015, which provides the same display.

Consider next the selective store function. It may be desirable to not look at every address, but only at those corresponding to instruction fetch cycles. The operator can do this by using the analyzer's display qualifier feature. Looking back at the sample program, Figure 8-34(b), observe that the subroutine is 14 instructions in length, with each instruction in the subroutine requiring at least two memory locations. In turn, the operator cannot view the entire subroutine on the 16-word display in Figure 8-33. By qualifying the display on the two-cycle control bits, it is possible to look at only addresses corresponding to instruction fetch cycles. The operator does

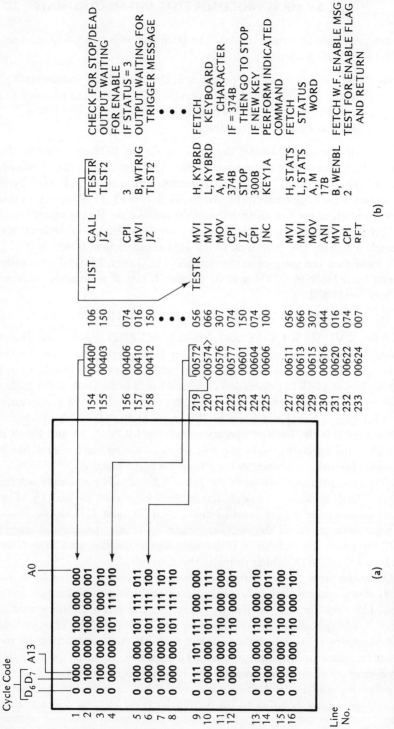

Figure 8-33. System response to CALL instruction. (Courtesy, Hewlett-Packard)

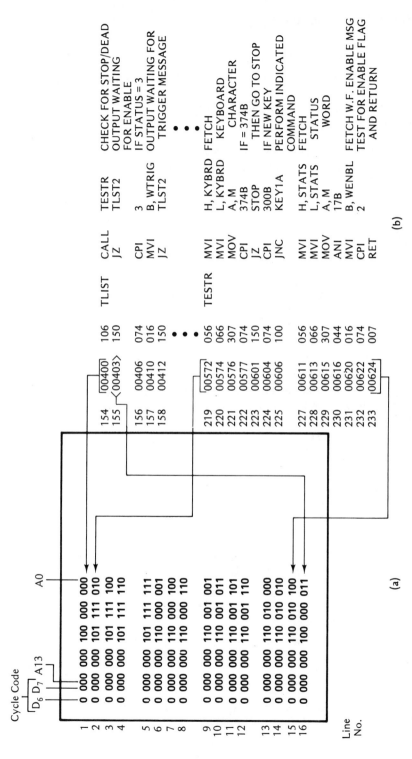

Figure 8-34. Qualified display showing the ability to selectively display only one desired addressed data. (*Courtesy*, Hewlett-Packard)

228 DIGITAL MEASUREMENTS

this in the following manner:

1. Connect Q1 and Q0 probes to monitor cycle control bits D6 and D7.
2. Set DSPLY/TRIG pushbutton to DSPLY.
3. Set Q1 and Q0 switch to LO.

The state display shown in Figure 8-34(a) is then obtained. Bits 15 and 14 are both zero for every displayed address, indicating that each displayed address represents an instruction fetch. Comparing the table display with the program listing reveals that line 1 is the address of the CALL instruction, lines 2 through 15 denote the subroutine, and line 16 is the return to the main program. Thus, the operator has an overview of the entire subroutine.

Next, consider the map. If a tabular display is not presented in the foregoing procedure, it means that the system did not access the selected address, and the No Trigger light will be on. To find where the system is residing in the program, switch to Map (Figure 8-35). Using the Trigger Word switches, move the cursor to encircle one of the dots on the screen. Switch to Expand and make the final positioning of the cursor. The No Trigger light will then go out, and switching back to Table A displays the 16 addresses around that point.

When program deviations are found, the reason may be as simple as a program error or as complicated as a hardware failure on the DATA/CONTROL BUS, or other command lines. It is helpful in this situation to employ additional input channels by combining the 1600A and 1607A analyzers. This expands the trigger capability to 32 bits wide, allowing the 14-bit addresses, 8 bits of data, and up to ten other command signals to be viewed simultaneously. Observe the sample program in Figure 8-36(b). By

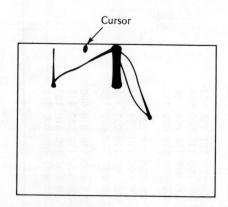

Figure 8-35. Map display shows entire system activity. (*Courtesy,* Hewlett-Packard)

```
0 000 000 100 000 000
0 100 000 100 000 001
0 100 000 100 000 010          01 000 110
0 000 000 101 111 010          01 111 010
                               00 000 001
0 100 000 101 111 011          00 101 110
0 000 000 101 111 100
0 100 000 101 111 101          00 111 011
0 000 000 101 111 110          00 110 110
                               11 000 000
                               11 000 111
0 111 101 111 000 000
0 000 000 101 111 111          01 111 010
0 100 000 110 000 000          11 111 100
0 000 000 110 000 000          11 111 100
0 000 000 110 000 001          01 101 000

0 100 000 110 000 010          01 101 000
0 000 000 110 000 011          00 000 001
0 000 000 110 000 100          00 111 100
0 100 000 110 000 101          11 000 000
```

154		TLIST	CALL	TESTR	CHECK FOR STOP/DEA
155			JZ	TLST2	OUTPUT WAITING
					FOR ENABLE
156	00400	106	CPI	3	IF STATUS = 3
157	00403	150	MVI	B,WTRIG	OUTPUT WAITING FOR
158	00410	074	JZ	TLST2	TRIGGER MESSAGE
	00410	016			
	00412	150			
		•••			
219	00572	056	TESTR MVI	H,KYBRD	FETCH
220	00574	066	MVI	L,KYBRD	KEYBOARD
221	00576	307	MOV	A,M	CHARACTER
222	00577	074	CPI	374B	IF = 374B
223	00601	150	JZ	STOP	THEN GO TO STOP
224	00604	074	CPI	300B	IF NEW KEY
225	00606	100	JNC	KEY1A	PERFORM INDICATED
					COMMAND
227	00611	056	MVI	H,STATS	FETCH
228	00613	066	MVI	L,STATS	STATUS
229	00615	307	MOV	A,M	WORD
230	00616	044	ANI	17B	
231	00620	016	MVI	B,WENBL	FETCH W.F. ENABLE MSG
232	00622	074	CPI	2	TEST FOR ENABLE FLAG
233	00624	007	RET		AND RETURN

Figure 8-36. System response to CALL on address, data, and control lines. (*Courtesy*, Hewlett-Packard)

230 DIGITAL MEASUREMENTS

displaying both address and data, it is now possible to confirm exact system operation with respect to the CALL instruction. Looking at line 1 of the state display in Figure 8-36(a), observe that bits 15 and 14 of the left-hand table are 00, indicating that the 8 bits of displayed data represent an operation code. Thus, 01 000 110 is the code for the CALL instruction. The second and third byte of a CALL instruction should be the lower and upper address bits, respectively, of the subroutine being called. Examination of the address in the fourth line reveals that, indeed, the data bytes of lines 3 and 2 (00 000 001 and 01 111 010) have been combined to form the subroutine address (00 000 101 111 010). In a similar manner, each line of the display can be examined to reveal exact program operation.

Frequency Measurement

Frequency (repetition-rate) measurements can be made at any point in a digital system by means of an oscilloscope with triggered sweeps and a calibrated time base. The procedure is simplified, however, if a digital frequency counter is employed, as illustrated in Figure 8-37. A counter is direct-reading; it eliminates the necessity for screen-pattern evaluation and conversion of time periods into frequency values. For these reasons, a counter measurement is also more reliable, particularly when utilized by comparatively unskilled personnel.

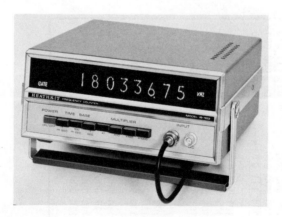

Figure 8-37. A digital frequency counter. (*Courtesy*, Heath Co.)

Review Questions

1. Distinguish between the time-frequency domain and the data domain.
2. Are electrical and functional analyses separable? Why?
3. Describe a logic state analyzer.
4. What is the difference between a bit and a byte?
5. How many bytes are there in a digital word?
6. Describe the meaning of setup time.
7. State two typical logic levels.
8. Discuss the nature of a truth table.
9. What is the function of a logic probe?
10. Describe a glitch.
11. Discuss the difference between alternate-channel and chopped-mode displays.
12. Distinguish between positive delay and negative delay.
13. Define a race condition.
14. Briefly describe system response to a test routine.
15. What is a digital map display?

appendix i

Resistor Color Code

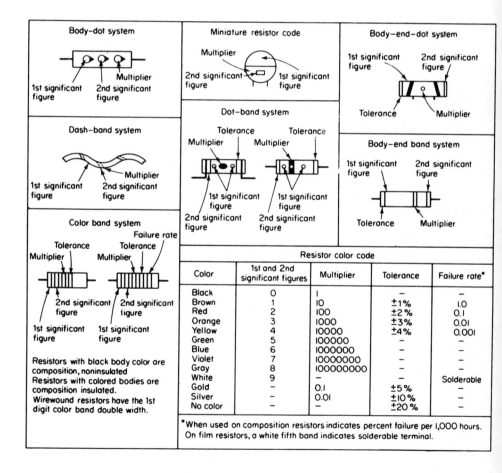

appendix ii

Capacitor Color Code

Molded mica capacitor codes (capacitance given in MMF)

Color	Digit	Multiplier	Tolerance	Class or characteristic
Black	0	1	20 %	A
Brown	1	10	1 %	B
Red	2	100	2 %	C
Orange	3	1000	3 %	D
Yellow	4	10000	–	E
Green	5		5 % (ELA)	F (JAN)
Blue	6			G (JAN)
Violet	7			
Gray	8			I (ELA)
White	9			J (ELA)
Gold		0.1	5 % (JAN)	
Silver		0.01	10 %	

Class or characteristic denotes specifications of design involving Q factors, temperature coefficients, and production test requirements.
All axial lead mica capacitors have a voltage rating of 300, 500, or 1000 volts, for 4.0 MMF whichever is greater.

Molded paper capacitor codes (capacitance given in MMF)

Color	Digit	Multiplier	Tolerance
Black	0	1	20 %
Brown	1	10	
Red	2	100	
Orange	3	1000	
Yellow	4	10000	
Green	5	100000	5 %
Blue	6	1000000	
Violet	7		
Gray	8		
White	9		10 %
Gold			5 %
Silver			10 %
No color			20 %

Indicates outer foil. May be on either end. May also be indicated by other methods such as typographical marking or black strip.

Molded paper tubular: 1st, 2nd significant figures; Multiplier; Tolerance.

Molded — insulated axial lead ceramics: 1st, 2nd significant figures; Multiplier; Tolerance; Temperature coefficient.

Typographically marked ceramics: Temperature coefficient; Capacity; Tolerance.

JAN letter	Tolerance	
	10 MMF or less	Over 10 MMF
C	±0.25 MMF	
D	±0.6 MMF	
F	±1.0 MMF	±1 %
G	±2.0 MMF	±2 %
J		±5 %
K		±10 %
M		±20 %

Extended range T.C. tubular ceramics: 1st, 2nd significant figures; Multiplier; Tolerance; Temp. coeff. multiplier; T.C. significant figure.

Color band system: 1st, 2nd significant figures; Multiplier; Tolerance.

Resistors with black body color are composition, non insulated. Resistors with colored bodies are composition, insulated. Wire-wound resistors have the 1st digit color band double width.

Resistor codes (resistance given in ohms)

Color	Digit	Multiplier	Tolerance
Black	0	1	±2 %
Brown	1	10	±1 %
Red	2	100	±2 %
Orange	3	1000	±3 %*
Yellow	4	10000	GMV*
Green	5	100000	±5 %
Blue	6	1000000	±8 %*
Violet	7	10000000	±12 1/2 %*
Gray	8	0.01 (ELA alternate)	±30 %*
White	9	0.1 (ELA alternate)	±10 % (ELA alternate)
Gold		0.1 (JAN and ELA preferred)	±5 % (JAN and ELA preferred)
Silver		0.01 (JAN and ELA preferred)	±10 % (JAN and ELA preferred)
No color			±20 %

Extended range T.C. tubular ceramics: Tolerance; 1st, 2nd significant; Multiplier.

Body-end band system: 1st, 2nd significant figures; Tolerance; Multiplier.

*GMV = guaranteed minimum value, or −0, 100 % tolerance.
±3, 6, 12 1/2, and 30 % are ASA 40, 20, 10, and 5 step tolerances.

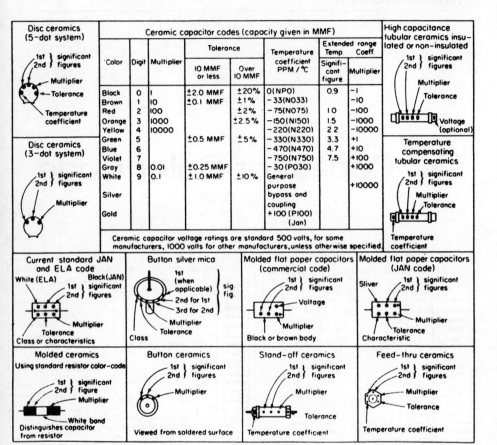

appendix iii

Basic Electrical Circuit Equations

Ohm's Law for Direct Current

$$I = \frac{E}{R}, R = \frac{E}{I}, E = I \times R$$

where, I = amperes,

E = volts,

R = resistance.

Resistances in Series

$$R_{(Total)} = R_1 + R_2 + R_3 + \text{etc.}$$

Resistances in Parallel

$$R_{(Total)} = \frac{1}{\frac{1}{R_1} + \frac{1}{R_2} + \frac{1}{R_3} \text{ etc.}}$$

Resonant Frequency

$$f = \frac{1}{2\pi\sqrt{LC}}$$

where, f = frequency in hertz,

L = inductance of the circuit in henrys,

C = capacitance of the circuit in farads.

Inductive Reactance

$$X_L = 2\pi fL$$

where, X_L = inductive reactance,

L = inductance in henrys.

Capacitive Reactance

$$X_C = \frac{1}{2\pi fC}$$

where, X_C = capacitive reactance,

C = capacitance in farads.

Ohm's Law for Alternating Current

$$I = \frac{E}{Z} \text{ or } I = \frac{E}{\sqrt{R^2 + \left[(2\pi fL) - \left(\frac{1}{2\pi fC}\right)\right]^2}}$$

where, I = current,

E = voltage,

Z = impedance (total of all oppositions).

Capacitors in Parallel

$$C_{(Total)} = C_1 + C_2 + C_3 \text{ etc.}$$

Capacitors in Series

$$C_{(Total)} = \frac{1}{\frac{1}{C_1} + \frac{1}{C_2} + \frac{1}{C_3} \text{ etc.}}$$

Resonance

$$2\pi fL = \frac{1}{2\pi fC}$$

where, $2\pi fL$ = inductive reactance,

$2\pi fC$ = capacitive reactance.

$$f = \frac{1}{2\pi\sqrt{LC}} \text{, or } L = \frac{1}{(2\pi f)^2 C} \text{, or } C = \frac{1}{(2\pi f)^2 L}$$

where, f = resonance frequency in *hertz*,

L = inductance in *henrys*,

C = capacitance in *farads*.

Impedance of Resistance, Capacitance, and Inductance in Series

$$Z = \sqrt{R^2 + \left(2\pi fL - \frac{1}{2\pi fC}\right)^2}$$

Impedance of Resistance, Capacitance, and Inductance in Parallel

$$Z = \frac{RX_L X_C}{\sqrt{X_L^2 X_C^2 + R^2(X_L - X_C)^2}} \text{ ohms}$$

Impedance Relations in Series and Parallel Resonant Circuits

Parallel

$$Z = \frac{2\pi fL}{4\pi^2 f^2 LC - 1}$$

At resonance:
$$Z = Q2\pi fL$$

Series

$$Z = \sqrt{\left(2\pi fL - \frac{1}{2\pi fC}\right)^2 + R^2}$$

At resonance:
$$Z = R$$

where Q is the "factor of merit" of the coil $= \dfrac{2\pi fL}{R}$

Power in AC Circuits

$$W = E \times I \times \frac{R}{Z}, \text{ or } E \times I \times \text{cosine } \phi$$

where, $W =$ power in *watts*

and $\dfrac{R}{Z}$ is called the *Power Factor*.

$$PF = \frac{\text{true power}}{\text{apparent power}} = \frac{I \times R}{E}$$

Index

Abnormal distribution, 81
Absolute:
   error, 40
   units, 3
AC:
   component, 69
   converter, 91
   resistance, 43, 50
   voltage measurements, 87
Accumulator, 214
Accuracy, 27
   rating, 10, 13, 26
   specification, 14
Address:
   flow, 212
   line, 212
   stack, 225
AFC pulses, 148
AGC pulses, 148
Alternate mode, 205
Ampere unit, 2
Amplifier:
   high fidelity, 172
   horizontal, 110
   vertical, 110
Amplitude compensation, 137
Analog:
   current meter, 4
   display, 204
   ohmmeter, 40
   voltmeter, 1
AND gate, 208
AND/OR gate, 196
Anechoic room, 188
Apparent distortion, 173
Arc of error, 40
Arcing, 161
Attenuator, 10

Audio:
   amplifiers, 171
   characteristics, 172
   composite signal, 183
   frequencies, 20
   measurements, 165
   oscillator, 180
   power bandwidth, 171
   units, 184
   voltmeter, 93
Average:
   response, 13
   value, 37

Balance, null, 58
Ballast resistor, 52, 158
Bandpass filter, 174
Bandwidth, 165
Base circuit, 70
Base-line level, 115
Beta, 134
Bias:
   signal-developed, 83
   voltage, 75
Binary:
   digits, 191
   form, 191, 195
   words, 191
Bits, 191
Black compression, 142
Blanking pulses, 148
Blocking capacitor, 89
Bogie values, 149
Breaker points, 153
Bridge:
   Hay, 181
   impedance, 179

242 INDEX

Bridge: (*Contd.*)
   inductance, 179
   Maxwell, 279
   null balance, 58
      indication, 179
   RCL, 148
British Association, 3
   units, 3
Bureau of Standards, 14
Buzz, sync, 160
Byte, 193, 220

Cadmium cell, 4
Calculator application, 15
CALL instruction, 222
Capacitance:
   bridge, 181
   values, 180
CD system, 158
Channel resistance, 61
Chopped mode, 205
Circuit:
   disturbance, 34
   loading, 32, 65, 112
   malfunctions, 82
CMOS, 198
CM-RAM, 211
CM-ROM, 211
Class A, 78
Clock:
   cycles, 193
      delay, 193
   rate, 192
   signal, 191
Coincidence, 204
Common:
   base circuit, 70
   collector circuit, 70
   drain circuit, 73
   emitter, 68
   gate circuit, 73
   ground, 68
   source circuit, 73
Compensation, probe, 113
Complex:
   tone, 187
   waveforms, 39
Compression, 142
Concrete standards, 3
Conditional jump, 214
Configuration:
   CB, 71
   CC, 71
   CE, 71
Conventional current, 68
Converters, 90
CPU, 208
Cross checks, 31
Crossover:
   distortion, 173
   network, 182
CRT, 176
Current:
   burden, 32, 89

Current: (*Contd.*)
   flow, 68
      reverse, 80
   gain, 134
   leads, 35
   measurements, 79
   tracer, 196
Curve tracer, 131, 134
Cutoff frequency, 141

Daniell cell, 4
Damped exponential, 119
Damping time, 107
Data domain, 191
   analyzer, 20
   display, 193
Data word, 195
dB:
   absolute values, 102
   correction factor, 101
   reference value, 99
   relative values, 102
dBm, 102
   chart, 103
DC:
   component, 13, 89
   resistance, 43, 50
   voltage, 5
      distribution, 68, 75, 77
Decade resistor box, 28
Decay time, 112
Decibel:
   meter, 96
   measurements, 96
   table, 98
Decoder, 183
Delay:
   line, 107
   section, 107
   time, 137
Demodulation, 129, 175
Demodulator probe, 129
Demultiplexed data, 216
Digital:
   addresses, 191
   bits, 191
   calculator, 16
   checkout, 193
   delay, 195, 206
   display, 13
   equipment frequencies, 20
   events, 191
   information, 192, 193
   instructions, 191
   measurements, 192
   multimeter, 45, 90
   pattern, 206
   probe, 196
   pulses, 198
   signal, 192, 195
   states, 197
   storage, 195
   triggering, 206
   voltmeter, 1, 13
   words, 191

INDEX 243

Direct-coupled circuitry, 83
Disc memory, 206
Display:
    clock signal, 192
    time-frequency, 192
Distorted waveforms, 160
DMM, 90
DTL logic, 197
Dual trace display, 195
Duty cycle, 107
DVM, 13
Dynamic impedance, 133

ECL logic, 200
Edge:
    leading, 110, 114
    pointer, 29
Electrical:
    analysis, 191
    measurements, 1
    parameters, 192
    units, 1, 3
Electrodynamometer, 87
Electrometer, 1
Electromotive force, 4
Electron current, 68
EMF, 4
Emitter coupled logic, 200
Envelope, wave, 129
Environmental factors, 32
Equalization, 137
Equalizers, 140
Equipment frequencies, 20
Error:
    absolute, 40
    arc of, 40
    band, 115
    computation, 36
    indication, 34
    offset, 28
    parallax, 30
    percent:
        of full scale, 10
        of reading, 10
    probable, 127
    turnover, 91
    waveform, 91, 97
Events, random, 207
Evolute waveform, 111
Expansion, pulse, 114
Experimental procedure, 25
Exponential:
    damped, 119
    wave, 119
External trigger, 204

Factors:
    dB correction, 101
    environmental, 32
    precision, 30
Fall time, 110, 115
Fast pulse measurements, 110
FET, 72
Fetch cycles, 227

Field:
    effect transistor, 72
    strength meter, 94
Filter:
    bandpass, 74
    tunable, 94
Firing pattern, 155
Flashing:
    intermittent, 161
    random, 161
Flip-flop, JK, 202
Floating base, 81
Flow:
    address, 212
    of current, 68
FM oscillator, 175
Form, binary, 191
Forward bias, 75
Frequencies:
    audio, 20
    equipment, 20
    hypersonic, 20
    operating, 127
    radio, 20
    sideband, 174
    ultrasonic, 20
    video, 21
    zero, 20
Frequency:
    characteristics, 165
    counter, 191
    modulation, 175
    response, 35, 90, 165
        curves, 165
Front-to-back ratio, 44
Full scale:
    current, 18
    voltage, 18
Functional:
    analysis, 191
    display, 191
    measurement, 193
    relationships, 195

Gage, strain, 44, 57
Gain:
    current, 134
    voltage, 165
Galvanometer, 58, 180
Gate:
    AND, 208
    AND/OR, 196
    circuit, 173
    NAND, 214
Gauss, Karl Friedrich, 2
General Conference, 5
Generator:
    audio, 56
    UHF, 21
    VHF, 21
Glitch, 199
Ground, common, 68
Grounded:
    base circuit, 70
    collector circuit, 70

244 INDEX

Grounded: (*Contd.*)
   drain circuit, 71
   gate circuit, 71
   source circuit, 71

h.a.d., 136
Half:
   amplitude duration, 136
   digit, 13
   sine wave, 36
   wave rectifier, 92
Hangover effect, 188
Harmonic:
   distortion, 165
      analyzer, 94, 172
   second, 131
   third, 31
Hay bridge, 181
HDM, 172
High:
   fidelity, 110, 167
   frequency:
      measurements, 124
      response, 35
   impedance:
      circuitry, 112
      voltmeter, 35
   logic level, 192, 196
   nodes, 196
   power ohmmeter, 44
   speed sweep, 113
   threshold logic, 198
   voltage probe, 83
HiNIL logic, 198
Hole current, 68
Horizontal drive pulses, 148
HTL, 198
Hum voltage, 173
Hypersonic frequencies, 20

IC:
   inputs, 199
   outputs, 196
IEEE standards, 5
IF alignment, 139
Ignition:
   analyzer, 153
   waveforms, 153
IM, 176
Impedance:
   bridge, 179
   components, 179
   dynamic, 133
   incremental, 133
   input, 165
   negative, 165
   output, 133
In-circuit:
   measurement, 45, 46
   resistance check, 60
Incremental resistance, 44
Indication error, 34
Inductance bridge, 179

Input:
   characteristic, 170
   impedance, 165
   port, 167
   resistance, 32, 54
Input/output characteristics, 168
Instructions, digital, 210
Instrument:
   circuitry, 17, 38
   rectifiers, 32, 90
Intangible resistance, 44
Integrated circuits, 196
Intensified interval, 203
Intercarrier sound, 150
Interface, 214
Intermediate frequencies, 20
Intermittent flashing, 161
Intermodulation:
   analyzer, 94
   distortion, 165
      transient, 177
   test signal, 177
Internal resistance, 54, 59
International:
   ampere, 4
   committee, 4
   system, 5
   volt, 4
Isolating resistor, 35

JCN instruction, 216
   addresses, 215
   checkout, 216
JFET, 72
   N type, 72
   P type, 72
Jittery pattern, 161
JK flip-flop, 202
   data transfer, 202
   timing levels, 202
Joule's law, 19
Jump, conditional, 216
Junction:
   field effect transistor, 72
   leakage, 79
   open-circuited, 79
   resistance, 44
   short-circuited, 81
   transistor, 68

Key signals, 199
Keyed rainbow signal, 144
Knife-edge pointer, 29
Kohlrausch, Friedrich Georg, 2

L channel, 184
   separation, 183
LDR, 44, 53, 202
Leading edge, 110, 114
   rise time, 115
Leakage resistance, 44, 79
   megohmmeter, 56
Light-dependent resistance, 44

Linear:
  log scale, 9
  phase sweep, 144
  resistance, 44
  units, 9
Linearity measurements, 124
Lissajous pattern, 122
Lobe development, 140
Lobes, 138
Logarithmic units, 9
Logic:
  clip, 191
  CMOS, 198
  comparator, 191
  current tracer, 196
  DTL, 197
  ECL, 200
  high, 192, 196
  HiNIL, 198
  HTL, 198
  low, 192, 196
  MOS, 198
  multifamily, 197
  NMOS, 198
  packages, 196
  probe, 191, 196
  RTL, 198
  state test, 193
  TTL, 196
London Conference, 4
Low:
  capacitance probe, 112
  frequency response, 35
  impedance circuitry, 112
  level tests, 93
  nodes, 196
  power ohmmeter, 44

Malfunctions, circuit, 82
Map display, 214
Marker adder, 129
Marker placements, 204
Markers, 129
Maxwell bridge, 179
Measurement:
  errors, 25
  levels, 193
Megger, 56
Megohmmeter, 56
Mel, 186
Memory, digital, 210
Mercury battery, 7
Meter:
  decibel, 96
  movement, 10, 18
  response, 39
  VU, 185
Microprocessor, 203
Mirrored scale, 29
Modulation, 175
MOSFET, 72
MOS logic, 198
Multifamily logic, 197
Multiline signals, 195

Multimeter, 17, 45, 90
  calibrator, 8
Multiplexed data, 216
Multiplier:
  cartridge, 83
  resistor, 18, 32, 83
Music power, 165

NAND gate, 214
National Bureau of Standards, 14
Negative:
  impedance, 104
  peak value, 87
  resistance, 43, 51
  temperature coefficient, 44
Network, crossover, 182
New key, digital, 226
NMOS, 198
Node, 199
Nodes, 196
Noise:
  deciphering, 192
  glitch, 199
  immunity, 200
  output, 165, 173
  waveform, 194
Nonlinear:
  characteristic, 49, 142
  reactance, 105
  resistance, 43, 49
Nonsinusoidal, 32
Normal curve, 26
NORM operation, 214
NOT AND gate, 214
NPN transistor, 68
NTSC, 144
N-type FET, 72
Null:
  balance, 58
  indication, 179
Numbers:
  binary, 197
  terminal, 210

Observational error, 30
Offset:
  color subcarrier, 144
  error, 28
Ohm, George Simon, 2
Ohmmeter, 140
  high power, 40
  low power, 40
Ohm's law, 19
Ohms per volt, 32
Ohm unit, 2
One-half digit, 13
Open circuit, 79
Operating frequencies, 127
Operational stack, 16
Oscillations, 156
Oscillator:
  action, 36
  audio, 180
  UHF, 136

Oscilloscope:
    data domain, 20, 193
    dual trace, 123
    tests, 107
Output:
    impedance, 133
    level meter, 166
    noise, 165, 173
    power, 165, 172
    resistance, 54
Overdrive, 69
Overdriven amplifier, 147
Overrange, 10
    digit, 13
Overshoot, 115, 137
    lobes, 140

Parade pattern, 155
Parallax error, 30
Parallel data stream, 206
Partial digit, 13
Peak:
    current, 51
    response, 13
    to peak voltage, 115
Phase:
    abnormalities, 140
    equalization, 137
    equalizers, 140
    irregularities, 137
    measurement, 121
    shifter, 121
Phon, 185
Photovoltaic cell, 59
Pickup assembly, 158
Pip (marker), 126
PNP transistor, 68
Positive:
    peak value, 87
    resistance, 43, 51
    temperature coefficient, 44
Post injection markers, 128
Power:
    bandwidth, 165
    output, 165, 172
    sink, 105
    source, 105
Preamplifier, 167
Precision:
    definition, 27
    factors, 30
Predistortion, 137
Primary:
    standard, 3
    waveform, 157
Printed circuit, 19
Probable error, 27
Probe compensation, 113
Probing, 202
Program:
    deviations, 229
    listing, 214
P-type FET, 72
Pulsating DC, 89

Pulse:
    amplitude, 115
    distortion, 147
    expansion, 114
    generator, 191
    single-shot, 196
    test, 178
    trains, 199
    width, 107, 115, 196
Pulser, 191, 197
Pulses:
    AFC, 148
    AGC, 148
    blanking, 148
    sync-buzz, 148
    transfer, 199

Qualified display, 228
Qualifier, 218
Q value, 119, 180

Race condition, 200
Radio frequencies, 20
RAM, 210
    chip, 211
Ramp, 124
Random:
    access memory, 210
    events, 207
    flashing, 161
Range switch, 65
R channel separation, 183
RCL bridge, 180
Reactance variation, 105
Reactive components, 54
Read Only Memory, 210
Readout, 203
    accuracy, 204
Real time, 218
Rectified waveform, 37
Rectifier probe, 92
Reference volume, 185
Reflection time, 117
Register, MPX, 209
Relay logic, 200
Repetition rate, 107
Reset signal, 199, 210
Resistance:
    chart, 48
    standard, 3
Resonant system, damped, 120
Response curve, 125
    tolerance, 126
Reverse current flow, 44, 80
Ringing, 115, 141
    lobe development, 140
Rise time, 107, 110, 115
RMS:
    units, 87
    value, 38
ROM, 210
    bank, 211
    chip, 212
Root mean square, 13

INDEX 247

Scale:
  factor, 97
  plate, 10
  resolution, 40
Second:
  harmonic, 31
  waveform, 157
Secondary standard, 3, 8
Sectional interactions, 83
Sensitivity, 32
Separation test, 183
Serial-A data, 206
Settling time, 115
Setup time, 195
Shelf life, 8
Shift signal, 199
Sideband frequencies, 174
Sidelock signal, 144
Siemen, 6
Signal:
  developed bias, 83
  generator, 22
  nodes, 191
  reset, 199
Sin^2 waveform, 136
Sine waveform, 88
Single-shot pulse, 196
SI units, 5
Slew rate, 115
Solid-state:
  multimeter, 12
  voltmeter, 35
Spark zone, 159
Spectrum analyzer, 173
Spikes, 192
Square wave, 110
  response, 110
  voltage, 36
Staircase voltage, 132, 141
Standard:
  cells, 5
  resistor, 2
Start signal, 199
State display, 229
Statistics, 25
Step:
  function, 118
  response, 139
Stereo decoder, 183
  separation, 183
Strain gage, 44, 57
Stretching distortion, 173
Subroutine, 225, 229
Superimposed pattern, 155
Surge, 153
Sweep:
  frequency test, 125
  generator, 125
  response curve, 124, 125
  tolerances, 126
Switching circuit, 52
Sync-buzz pulses, 148

Tangent galvanometer, 3
Tangible resistance, 44

Television frequencies, 20
Temperature:
  coefficient, 13
  variation, 13
Tesla, 6
Thermal resistance, 44
Thermistor resistance, 52
Thermocouple, 32, 95
  construction, 96
Third harmonic, 31
Tilt, 115, 179
TIM, 177
Time, 218
  base, 107
    triggered, 110
  constant, 117
  frequency domain, 191
  reflection, 117
Timing:
  diagram, 222
  levels, 203
  pulses, 153
Tolerance, 47, 82
Tone burst, 187
T pulse, 136
Transfer:
  characteristic, 47, 78, 141, 173
  pulses, 199
  signal, 199
Transient:
  distortion, 113
  response, 110, 135, 165
Transistor:
  measurements, 131
  voltmeter, 20, 92
Transit time, 116
Transmission channel, 140
Triggered sweep, 108, 118
Triggering, 202
Trigger point, 195
True rms, 13
  instruments, 87
Truth tables, 196
Tunable filter, 94
Tuned AC voltmeter, 94
Tunnel diode, 51
Turn-on action, 45
Turnover error, 91
TVM, 92
Two-tone signal, 177

UHF:
  generator, 21
  oscillator, 36
Ultrasonic frequencies, 20
Undershoot, 137
Unipolar transistor, 71
Universal system, 2
Unsaturated cell, 5
Unsymmetrical waveform, 92

Valley current, 51
Vectorgram, 143
Vectorscope, 144

248 INDEX

Vertical interval test signal, 135, 142
VHF:
   alignment, 129
   generator, 21, 125
Video:
   carrier, 129
   detector, 129
   frequencies, 21
   frequency spectrum, 137
   marker, 127
Visual alignment, 124
VITS, 135, 142
Vocal waveforms, 178
Volta, 2
Voltage:
   gain, 165
   leads, 35
   peak-to-peak, 115
   swing, 192
Voltaic:
   cell, 4
   pile, 2
Voltmeter:
   input resistance, 33
   sensitivity, 32
Volt unit, 2
Volume units, 184
VOM, 90
VTVM, 65
VU, 184
   meter, 185

Wave:
   envelope, 129
   two-tone, 177
Waveform:
   amplitude, 124
   analysis, 111, 200
   characteristics, 147
   decay time, 112
   detail, 114
   dissector, 175
   error, 91, 97
   $\sin^2$, 136

Waveform: (*Contd.*)
   storage, 111
   sweep alignment, 125
   visual alignment, 127
Waveforms:
   damped exponential, 119
   difference, 111
   distorted, 160
   evolute, 111
   ignition, 153
   involute, 111
   primary, 157
   product, 111
   pulse, 149
      single-shot, 196
   quotient, 111
   secondary, 157
   steady state, 152
   sum, 111
   sync
      buzz, 150
      train, 151
   transient, 152
   vocal, 178
Weber, Wilhelm Eduard, 2
Weston cell, 4
Wheatstone bridge, 58
White compression, 142
Window:
   digital events, 191
   pulse, 137
Wired-OR, 199
Word:
   digital, 193
   display, 202
   flow, 191
Working standard, 3

Zero:
   adjustment, 28
   frequency, 20
   temperature coefficient, 44
Zone, spark, 159